The shape of space

The shape of space

GRAHAM NERLICH

Professor of Philosophy
University of Adelaide, South Australia

CAMBRIDGE UNIVERSITY PRESS

Cambridge
London · New York · Melbourne

Published by the Syndics of the Cambridge University Press
The Pitt Building, Trumpington Street, Cambridge CB2 1RP
Bentley House, 200 Euston Road, London NW1 2DB
32 East 57th Street, New York, NY 10022, USA
296 Beaconsfield Parade, Middle Park, Melbourne 3206, Australia

Printed in Great Britain at the
University Printing House, Cambridge
(Euan Phillips, University Printer)

Library of Congress cataloguing in publication data
Nerlich, Graham, 1929–
The shape of space.
Bibliography: p.
Includes index
1. Space and time. 2. Relation (Philosophy)
3. Science–Philosophy. I. Title.
BD632. N45   114   75-44583
ISBN 0 521 21101 8

To Andrew, David and Stephen

Contents

Preface

It has been my aim in this book to revive and defend a theory generally regarded as moribund and defeated. Consequently, I owe most of my intellectual debt to those against whom I try to argue in these pages. Above all Hans Reichenbach's brilliant *Philosophy of Space and Time* has been the central work I wished to challenge and my model of philosophical style. My debt to the books and papers of Adolf Grünbaum is hardly less important. I am glad indeed to take this formal opportunity to express both my admiration of and my debt to these writers who are subjects of so much of the criticism in this book. Quite generally, indeed, I acknowledge a real debt to all those philosophers whose work it seemed important to examine critically in these pages.

This book is the fruit of courses I have given for a number of years in the philosophy departments of the University of Sydney and of Adelaide. Among many students whom it has been an education to teach, I hope it is not invidious to mention Michael Devitt, Larry Dwyer, Clifford Hooker and Ian Hunt as particularly helpful in early stages.

The whole of the typescript was read by Dr Ian Hinckfuss and Professor J. J. C. Smart (my first teacher in philosophy). Their comments and criticisms have been invaluable. Various parts of the book have been read by and discussed with Robert Farrell, Bas van Fraassen, Henry Krips, Robert Nola, Hugh Montgomery and Wal Suchting. I am grateful to my colleague Dr Peter Szekeres for his patience with many queries I have put to him about General Relativity. Of course, the responsibility for the errors and, perhaps, the perversity of the book is all my own.

Parts of the book were written in 1973 when I was Leverhulme Fellow at the University of Hong Kong and a visitor to the University of Auckland. I am grateful to my colleagues in both institutions for their very agreeable hospitality.

Chapter 7 first appeared under its present title in the *Journal of Philosophy* 70 (1973), 337–51.

I am grateful to Ms Iris Woolcock for the Index and to her and Ms Alvena Hall for some bibliographical work. Mrs A. Bartesaghi typed several drafts and read the proofs and has been an invaluable aid in many ways. Alice Turnbull helped with the figures.

Introduction

In this Introduction I want to give the aims and the plan of the book. Any reader may wish to know what it is about, how I saw my task in writing it and what knowledge I have assumed he will bring to the reading of it.

This book is about two philosophical questions: Is space something or nothing? If it is something, what is its structure? There is nothing new about the questions, of course. They have been, and they still are, energetically discussed. But the tide of philosophical opinion runs very strongly in favour of certain answers and it always has. The ancients spoke of space as The Void. This seems to have been meant as a subtle way of saying that space itself is nothing. There are only familiar things spatially related to each other in familiar ways. So began the long and illustrious tradition of answering the first, ontological, question in a relationist style. The bulk of what has been written on the subject is solidly relationist. I have set my face against the run of the tide in this book. I argue for an absolutist conclusion. Space is an entity in its own right – a real live thing in our ontology.

Relationism comes in degrees. The sternest stuff is the doctrine that the idea of space is nonsense – no idea at all. Only talk about material things and their relations can be understood. A milder view is that relationism is more plausible, neater, clearer than absolutism. Mildest of all is the claim that relationism is at least possible. We could say all we need to say, confining our statements to those about material things and their relations. But this implies no judgement about what it is best to do. It does not even claim that existing relationist theories actually work. I have no quarrel with this mildest opinion. But I do argue that relationism must appear laborious, contrived and counter-intuitive beside the absolute picture of space.

A main problem with an absolutist answer to the first question lies in understanding it. This takes us straight to the question about structure. I argue that to understand space as a thing in our ontology is to understand it as a thing that has shape. Its shape is its structure. A great deal of what follows is devoted to explaining in detail what it is for space to be shaped. It raises questions of topology, of projecture, affine and metrical geometry. Questions of intrinsic geometry, intrinsic shape and intrinsic curvature get to be critical ones. Yet understanding all this ought not to be a matter just of abstract geometrical reasoning. We need an imaginative, visual grasp of it all. So questions of visual geometry take their turn for our attention.

There are snares in this seemingly straightforward path to the absolutist's goal. They are snares of philosophical method. In much of the literature on the subject absolutist explanations are written off as mere conventions of language, without factual sense or content. There is no grasping the full import of the absolutist message without also grasping what the message is a reaction to. What the relationist emerges as arguing for is that space and the world of spatial things is less structured that it seems to be. This is not an ontological reduction, like reducing the existence of space to the existence of a collection of things related to one another. But structural reduction does have a kind of ontic bite. It says that real things are more vague, or amorphous, than we usually think. No doubt this is a step in the direction of banishing space. Space begins to look like an ontic spectre, a delusion or convention. The idea of structure comes in for some scrutiny therefore. The methodology of conventionalism is examined and rejected. This clears the way to a picture of space as having intrinsic shape and of geometrical explanation as an important part of our understanding of the world.

Early in the book, I argue that the shape of space globally gives us a geometrical explanation of a tangible, domestic puzzle: just how and why do left and right hands differ? This picture of what the shape of space might do for us sustains and motivates a lengthy examination in later chapters of what shape is for space and how we can defend shape as a genuine property of space. But in the final chapter of the book I try to illustrate the power of geometrical explanation to resolve much deeper and broader problems about how the physical world works. The examination of this begins with a problem classical in discussions about whether space is a real structural thing. It is the philosophical problem of the relativity of motion. I relate the question whether motion is relative to the question whether spacetime has a shape and argue for the superior power and richness of absolutist answers. This concludes the plan of the book.

I saw my task as weaving together four distinguishable threads of interest into an intelligible fabric. The overriding interest is to settle the ontological and structural questions, which I see as philosophical ones. But these cannot be settled without some grasp of various theories about what space is (or can be) like. So, much of the book is given up to informal, intuitive exposition of a wide range of ideas from geometry.

All of this is put to some philosophical use somewhere in the book even if not at the point of immediate exposition. The two threads of geometry and of philosophy (ontology and structure) interweave almost everywhere.

But relationists and I disagree. So there is a thread of criticism in the fabric, too. Much of this is dissent over the methods proper for settling questions of ontology and structure, rather than dispute directly about the answers to these questions. I have tried to confine this part of my argument just to Chapters 5 and 6, which therefore make a kind of interlude or detour from the main concern of the book. I think that the study of methodological problems is clearly motivated by the argument through which I lead up to it. But it is a separable aspect of the book. Similarly, there is a fourth thread of interest in the relativity theories of modern physics. They cast a great deal of light on the explanatory role which the shape of space might assume. Again, the exposition is everywhere informal and intuitive and always motivated by the relevance it will have for the ontological and structural questions. Relativity theory occurs only in the last chapter.

Clearly, then, this is a somewhat technical book. But it is aimed at the general philosophical reader. Its basic interest for me lies in the repercussions it has for our philosophical picture of what kinds of theory can properly be constructed; of how far the creative mind can outstrip bare sense experience. I see its wider implications as stretching right across the board of intellectual theorising. So I felt it crucial that the book should lie within the reach of a more general group of readers than those who specialise in the philosophy of science. I assume a good deal of maturity in a reader's ability to follow philosophical argument. But the main progress of the book can be followed without prior grasp of sophisticated mathematical or physical knowledge. I have tried to write as simply and clearly as possible. Perhaps the book is a little difficult, but I believe it can be thoroughly understood (for acceptance or rejection) with a little care and patience.

In detail, I assume little in mathematics beyond the basic concepts of high school algebra, elementary Euclidean geometry and classical physics. I also take it that the reader understands the basic concepts of calculus, that of a function and its derivative. No detailed knowledge of any of these subjects is presupposed. Where the notation of analysis etc. does appear, it is never crucial for an understanding of the matter in hand though, of course, it may help. There are exceptions to this in Chapters 3, 8 and 9 where more technical sections or paragraphs are clearly indicated and may be omitted without loss to the broad issues of the chapters. The treatment of the ideas of geometry and physics is always qualitative, pictorial and intuitive whenever a philosophical point depends on them. I treat the kinematics of Special Relativity at some length in a purely pictorial way. I supply simple models and diagrams for basic concepts of General Relativity, but my treatment of this theory is somewhat superficial and perfunctory. I go no further into it than seems necessary for a broad philosophical perspective.

If any reader wants a background to the more technical aspects of the subject, he can do no better than consult Adler (1958) on geometry and Feynman *et al.* (1963), Volume 1 on physics. Aleksandrov *et al.* (1964) is an immensely useful simple reference book on mathematics generally. These books are listed in the general bibliography. It has been my policy throughout to refer mainly to books that are both accessible and authoritative.

1 Space and systems of relations

1 Pure theories of reduction: Leibniz and Kant

Two great men stand out among those philosophers who have wished to exclude space from their list of things really in the world. They are Leibniz and Kant. Each man held views at some time which differed sharply from views he held at others. To take our bearings in the subject, let us begin by giving an outline of one theory which we owe to Leibniz's *Monadology*, another which we owe to Kant's *Critique of Pure Reason* and a third which we owe to Leibniz's letters to Samuel Clarke, a defender of Newton in the great seventeenth-century debate on mechanics.

Leibniz's argument in the *Monadology* is not epistemological but purely metaphysical. He was convinced that anything real must be a substance or its accident; that is, reality is comprised in things and intrinsic properties of things. Any apparent relation is either really an intrinsic aspect of a thing or it is nothing. Clearly enough, space and spatial relations are going to pose a problem.

Leibniz built a remarkably profound and elegant theory to show how to rid his ontology of space. Everything revolves round one nuclear idea: nothing mental can be spatial. Mind and mental attributes are a model for what is real but unextended. So Leibniz can regard the world as non-spatial if he can account for it all by regarding it as mental. This is an astonishingly bold and penetrating idea. It is worked out along these rough lines. Everything there is must be a monad, an unextended mind-like substance. Every monad has certain qualities which are to be understood on the model of perception. Monadic perceptions are relevant to spatial ideas for two reasons. First, they define a notional (ideal) relative 'position'. They are like photographs in that what appears on the film also defines a position from which the photograph was taken relative to the things that appear in it. The position defined does not depend in any way on where the photograph now is. Nor do the perceptions of monads need a place to define their notional position. But, second, perceptions are unlike photographs since perceptions are mental whereas photographs are spatial. The monads together with their perceptions define a logical ordering – a system of perceptions. This is space. But then space is not a real thing but only an ideal one, constructed out of monads. There could not be a real space across which these things are interrelated. To suppose there could would be to attribute to mental things properties which they cannot possibly have. (See Leibniz (1898) especially §§51–64. See Latta's footnotes to these sections.)

This is a sketch, not a portrait, of Leibniz's theory. What I mean to capture in it is a blueprint for what a fully reductive picture of space must be. For Leibniz, no objects are spatial and ultimately no real spatial relations hold among them. His theory is, therefore, a pure case of reduction and it is just its purity I want to emphasise. It contrasts with impure reductions in which space is reduced to extended objects with spatial relations among them. I call these impure reductions because spatial ideas remain, in some way, among the primitive ideas to which reduction leads. So long as that is the case it must remain seriously unclear just where this gets us. So purity is a gain. But Leibniz clearly paid a very high price in plausibility to make his system pure. The world does not seem to be composed of minds and nothing else. He evidently thought the preservation of substance and accident metaphysics was temptation enough for us to pay the price willingly. But I think the metaphysics in question has a strong appeal only if you believe that classical Aristotelian logic is all logic. Insofar as he did base his system on convictions about the nature of logic, not knowledge, his theory is pure metaphysics, not epistemology. However, Leibniz was not quite right about logic. Quantification theory now gives us a logic of relations which can be placed on foundations much more secure than Leibniz ever had for the syllogism. Further, it is a more powerful system. So it will be the cost of Leibniz's theory in plausibility that impresses us most. That remark is not intended to throw any stones at the consistency of Leibniz's system, nor at the purity of his reduction of space to objects.

We can find another case of pure reduction by glancing into Kant's philosophy. Though he admitted a debt to Leibniz, the deepest roots of Kant's thought are all epistemological. Kant believed that our experience was only of appearances, never of things themselves. There can be no reason to suppose that, in themselves, things really are as they appear to us. We can never get 'behind' appearances to check any resemblance between them and things as such. But, Kant argued, we do have reason to believe that, so far as appearances are spatial, they do not resemble the things that give rise to them. The reason is that we have necessary knowledge of geometry prior to any experience whatever;

this would be inexplicable unless space were something whose source lay wholly inside us. Kant was wrong, notoriously, in supposing that we actually do have any such knowledge. But what matters now is to understand the bold outlines of his theory, not to criticise it. The mechanics of the theory go something like this. Space is a pure form of perception which we bring to the matter of appearances and impose on it. Only by making appearances spatial can we experience them as objective, outer, and interrelated. But we make them spatial without the aid of empirical instruction from the appearances. Appearances come from things themselves and what they instruct in us is always *a posteriori*. So, if space is *a priori* it doesn't come from things. Things themselves are not spatial: they are not extended and they have no spatial relationships to each other. Space is purely a gift of the mind, not of things. It is ideal, not real. Things themselves are neither spatial nor spatially related (Kant (1953); (1961), pp. 65–101).

We have come to very much the same place as before, but along a different path. The point of discussing Kant was to see that there are indeed different paths. His spatial reduction is just as pure as that of Leibniz. It is also pretty equally costly in plausibility. For Kant things are forever beyond our ken and must remain so. We can grasp only their appearances. It is true that Kant here followed a long, distinguished, unfortunate tradition in philosophy. We will find reason, later, to abandon it. However, in both these examples, we rapidly scaled, by different tracks, a peak from which we glimpsed vast systems, comprehensive, elegant, pure and improbable. These heights are too dizzying and the air too rarefied for us to pitch our tents here unless we give up hope of finding another place to live. The metaphor is not a refutation, of course. But I am at pains to persuade the reader that the plausible reductions are impure and pure reductions are implausible. My real quarry is the plausible impure reduction that follows. I argue at length that this widely held theory about space will not work. The pure views of Leibniz and Kant are, I think, the obvious reductive alternatives to it. So I stress the tension between purity and plausibility. But let us look at this more ingratiating, less strenuous opinion.

2 Impure theories of reduction: outlines

In his famous correspondence with Samuel Clarke, Leibniz advances a view in the fifth letter, which is considerably weaker than the one we just saw. He writes as follows (Alexander (1956), pp. 69–70):

> 'I will here show, how Men come to form to themselves the Notion of Space. They consider that many things exist at once, and they observe in them a certain Order of Coexistence, according to which the relation of one thing to another is more or less simple. This Order is their Situation or Distance... Those which have such a Relation to those fixed Existents, as Others had to them before, have now the same Place which those others had. And That which comprehends all their Places, is called Space.'

Here Leibniz presents the idea of space as a construct out of extended bodies with spatial relations among them. He goes on to tell us what he thinks these remarks about space mean by drawing an analogy with genealogical trees.

> 'In like manner, as the Mind can fancy to itself an Order made up of Genealogical Lines, whose Bigness would consist only in the Number of Generations, wherein every Person would have his Place: and if to this one should add the Fiction of a Metempsychosis, and bring in the same Human Souls again; the Persons in those Lines might change Place; he who was a Father, or a Grand-father, might become a Son, or a Grand-son etc. And yet those Genealogical Places, Lines, and Spaces, though they should express real Truths, would only be Ideal Things.'

It is easy enough to see, from this, the general direction Leibniz wants to take. This time, there is no doubt about the plausibility of the idea, at least on the face of it. But the theory is impure however enthusiastically we regard it. Simply, there are spatial ideas among the primitives. But I put this worry to one side for now.[1] Before we ask for a more exact understanding of what Leibniz meant, let us see if we can find some Leibnizian echoes among modern writers on space and time.

P. F. Strawson has written some very interesting pages about space in his book *Individuals* (1959) and again in *The Bounds of Sense* (1966). Here are some quotations from the former book.

> 'There is no doubt that we have the idea of a single spatio-temporal system of material things; the idea of every material thing at any time being spatially related, in various ways at various times, to every other at every time. There is no doubt at all that this is our conceptual scheme' (p. 35).

> 'The system of spatio-temporal relations has a peculiar compre-

[1] It is discussed in §10 below.

hensiveness and pervasiveness, which qualify it uniquely to serve as the framework within which we can organize our individuating thought about particulars. Every particular either has its place in this system, or is of a kind the members of which cannot in general be identified except by reference to particulars of other kinds which have their place in it; and every particular which has its place in the system has a unique place there. There is no other system of relations between particulars of which all this is true' (p. 25).

'*For that framework is not something extraneous to the objects in reality of which we speak. If we ask what constitutes the framework*, we must look to those objects themselves, or some among them. But not every category of particular objects which we recognize is competent to constitute such a framework. The only objects which can constitute it are those which can confer upon it its own fundamental characteristics. That is to say, they must be three-dimensional objects with some endurance through time. They must also be accessible to such means of observation as we have; and, since those means are strictly limited in power, they must collectively have enough diversity, richness, stability and endurance to make possible and natural just that conception of a single unitary framework, which we possess. Of the categories of objects which we recognize, only those satisfy these requirements which are, or possess, material bodies – in a broad sense of the expression. *Material bodies constitute the framework*' [my italics] (p. 39).

Let us call this group of theories and opinions 'existential relationism'. What does this phrase convey? Relationist ideas, generally, tell us that what seem to be statements about space are really about relations among things. I reserve 'relativity' as a word to be applied to theories of rest and motion, which follows more or less established practice. Later, we will find a number of relational theories about space, each distinct from the others and from this. One theory is about spatial metric; another, quite distinct theory, is about topology. There could easily be others. They all need to be distinguished from the one before us. The present theory is about whether the existence of space is relational. So I choose the phrase 'existential relationism' which we might regard as a fine flower of our technical vocabulary. The theory may offer us either of two things. It may present us with a relational system as the object to which the noun 'space' refers. But it may offer just to parse away any sentence in which 'space' seems to need a direct referent. What we get, instead, is a sentence about things as spatially related. The claim is that nothing by way of power in explanation is lost in the change. Here, we will look at proposed referents for the noun 'space' and at the proposed reparsing in §8 of Chapter 2. Of course, these two tactics spill over into one another in practice. However, let me press on with explaining more exactly what the theory is.

3 A 'system of relations' theory: measurable qualities

How far is a genealogical tree really like space? What does someone mean by 'system' when he talks about a system of relations? We will not get a clear understanding of the Leibnizian view unless we ponder these questions pretty carefully. I want to answer them by looking into the structures we can uncover when we analyse a continuously measurable quality, preferably a multi-dimensional one, if we can find it. That will give us a much closer analogy with space than a genealogical tree can. Since Leibniz thinks that, at some point, we will take flight into the ideal, it will be useful to keep in mind whether and when we are dealing with tangibles. I will simplify the story by suggesting a more compliant world than we would find in practice, but the story remains realistic in principle. The world might have been exactly as I say.

I will pick weight as a simple example of a continuously measurable quality.[1] The example will play the part of suggesting systematic principles which can be extended to other cases generally. The field of the relation we are going to systematise, then, is the set of objects with weight. When we arrive at the system, we will see it to be an ordered set directly related to this one. Any clear explanation of systems of relations is unavoidably set theoretical, I believe. Clearly sets are abstract. Less clearly, they are real and part of our ontology.[2] We will not lose the heart of what Leibniz said if we speak of abstract, real sets rather than of what is ideal. This will certainly transform what he said, but the theory will be clearer and stronger for the change. We can see what the core of the issue is at the end of §6 when we have a wider view of what systems of relations are. So let us turn to the problem of explaining them.

A proper system of relations begins with a partial ordering; the relation, R (being heavier than, for example), partially orders the things

---

[1] I am indebted to N. R. Campbell (1920) and J. J. C. Smart (1959) here and elsewhere in this chapter.
[2] I follow Quine in this. See (1960), §55.

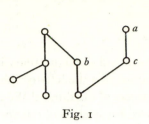

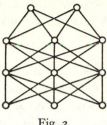

Fig. 1   Fig. 2

in its field if, and only if, it is transitive, asymmetrical and irreflexive in the set. That is, the schemata

$$(x)(y)(z)\,(Rxy \quad \& \quad Ryz \supset Rxz)$$
$$(x)(y)\,(Rxy \supset\, \sim Ryx)$$
$$(x)\,(\sim Rxx)$$

must be true in the case of the required substitution for R. But we also want the relation to be connected in the set strongly enough to provide for an adequate idea of equality in weight (or whatever else) which we might express by using the sign ' $\approx$ '. We can quickly see more or less what is needed if we look at the problem as Hasse diagrams illustrate it. These are drawings, in a certain sense to be explained, of ordered sets, which we can make use of now, and in criticism, later, as well. The elements of the set are drawn as circles and are connected by downward lines. The relation expressed by the open sentence 'x is joined by a downward line to y' is transitive and asymmetrical, so each diagram defines an actual partially ordered set whose members are the circles. It thus comes to represent any set isomorphic[1] with the one it defines. In fig. 1 we see a set whose connexions are too weak to give what we want. Look at the elements a, b and c. We could not say that equality under the relation, R, is the same as failure of R in both directions. For neither a nor c has R to b, nor does b have R to either of them. This might tempt us to say both that $b \approx a$ and that $b \approx c$. Yet it is obvious that '$a \approx b \approx c$' is false, since Rac. In short, the open sentence ' $\sim Rxy$ & $\sim Ryx$' does not yield R-equivalence classes in its field. The diagram of a measure connected set (i.e. one connected as we want) looks like fig. 2. Here two-way failure of the relation does yield satis-

---

[1] 'Isomorphic' literally means 'same shape'. Ordered sets are isomorphic roughly if there is a 1–1 correspondence between set members so that an object in one set has the same 'place' as its mate in the other set, under the ordering relation.

factory R-equivalence classes. Equivalence can now be defined by '$\sim Rxy$ & $\sim Ryx$'. We may now say:

$$(x)(y) \, (Rxy \lor Ryx \lor x \approx y)$$

Some analogies with genealogical trees and configurations in space are visible in the diagrams. However, notice that the left-to-right layout of them does not reflect any systematic aspect of the things.[1] We will look at problems of dimension later. Still, these are systems of relations of the kind Leibniz had in mind. It is clear, too, that the characteristics of the system are given to it by the weight characteristics of the things just as Strawson suggested.

4 Addition in a system of relations

This only provides for part of what is wanted. For, as Clarke pointed out to Leibniz (Alexander (1956), pp. 105–7), we can talk about the disposition of objects in space quantitatively. Object a is twice as far from b as c is. As I understand him, Clarke was arguing that space must be a substance if there are quantities of it. So far as I can see, Leibniz had no cogent answer to this point. It is a persuasive one. He did say, about genealogical trees, that we might talk of genealogical distance in terms of the number of generations. It is an interesting suggestion but not nearly good enough as it stands. However, it is certainly possible to speak of quantities in a measurable space. This does not require us to speak of quantities of the measure-space as a substance. How does that work?

Weighted objects can be added together. What is required for the present situation, though, is a little more complex than this might make it seem. Sometimes we mean by 'adding' simply 'mixing' or 'bringing together in the same region'. But we want an idea of adding which is also an idea of manipulating the objects just as the loose one is. By adding them, things will be combined into a third individual which counts for weighing purposes as an object of some heaviness. Nothing guarantees that this condition will actually be met, given only that there is a partial ordering relation with equality. We cannot for example add

---

[1] A little reflection on conditions like weak connexion will reveal that the field of our weight relation is vague. Does the moon have a weight? Does the earth? These problems are not instructive in the present context, however, so I have ignored them.

densities or temperatures by any manipulative operation.[1] But we can add weights in a satisfactory way. Bring the two objects together so that the weight of each acts through a common point or in a common area. This means that simply putting the things in a scale pan counts as adding their weights only if the pan is suspended from a single point on the beam. If it is suspended from two different points, one nearer the fulcrum of the balance than the other, the operation of adding weights will be more complex: we ought to ensure that one is on top of the other in the scale pan. Ordinarily, of course, things are so arranged in advance that these mechanical subtleties in the simple art of adding weights do not intrude on our procedure at all obviously.

The point of these more or less subtle precautions is to find a model of arithmetical addition in the concrete manipulative operation on things. We need not look, in detail, at the restrictions to be laid on this physical operation, **o**, so that it does become a model of $+$ among numbers. In many cases it is not at all difficult to meet them.

I do not mean by this that pure arithmetic is an uninterpreted calculus which implicitly defines what may count as adding. Pure arithmetic is an interpreted theory about numbers; perhaps this means that it is about certain sets and that '$a+b$' denotes the union of some pair of the sets. Whatever pure arithmetic is, exactly, applied arithmetic is best understood as a secondary theory in which the numbers are isomorphic with equivalence classes determined by $\approx$, with $+$ corresponding to **o** and $<$ to R. It would take us far afield to work these ideas out, however.[2]

Isomorphisms like the one we need can be set up in various ways. Simply choose a convenient object a to be the image of 1 (though, a's R-equivalence class is really the image of 1). Suppose there are two more objects b and c such that both $a \approx b$, and $a\mathbf{o}b \approx c$. Then the equivalence class of c is the image of 2 in the isomorphism, and so on. Pretty clearly, we will need at least rational numbers to carry the isomorphism through. There are bound to be objects b' and c' such that $b' \approx c'$ and $a \approx b'\mathbf{o}c'$. Thus $b' = c' = \frac{1}{2}$. And so on. We can easily find a use for the real numbers, too.

So now, in respect of quantity and continuity, as well as order, this system of relations is just like a space.

[1] For more detail see Campbell (1920) and Smart (1959).
[2] The isomorphism is idealised, of course. The set of objects with weight is not really closed under **o**.

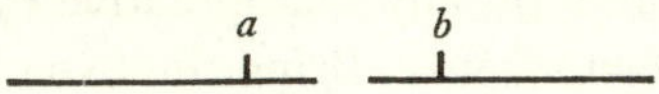

$$a \qquad\qquad b$$

Fig. 3

5 Dimensions in a system of relations

Let us look for more structure in a system of relations. At this point, a new example will be more helpful. The problem, now, is about the idea of dimensions.

Space has three dimensions. But weight 'space' has only one. So far as I know, there is no really satisfactory example of a many-dimensioned measurable quality-space. Yet if the world were just a little more obliging than it is, there could be one in a quite strict sense. My following discussion of dimensions fakes some conditions which smooth its path. Prompted by some ideas of P. F. Strawson, I choose sound for the next example.[1]

What decides the number of dimensions a continuum has? This question is about what sort of boundary (or cut) you need to divide the continuum into distinct parts. For example, if you take a Euclidean line and cut it at a point, then the two parts of the line become disconnected: you can choose two points *a* and *b* such that it is impossible to move from *a* to *b* without 'jumping' across the missing points, as in fig. 3.[2] Similarly, a surface can be divided into two parts by means of a curve. If the surface is infinite, then the curve can be a closed curve. But the removal of any countable set of points from the surface does not divide it, for we can always travel from *a* to *b* round any point, or between any pairs of points without leaving the surface. Further, in a three-dimensional space, neither points nor lines effectively divide it. Only surfaces can be boundaries in three-space. Again, they must be closed surfaces or envelopes if the space is finite. Points are assigned zero dimension. The number of dimensions of a space, then, can be defined as one more than the number of dimensions of an effective boundary in it.

Now we can shift our gaze from space to sounds. First of all, consider how sounds can form a system of relations with respect to their pitch. Without going into details, it is clear that we can partially order sounds in a scale of pitch and also speak of a sound being double the pitch of

---

[1] Another familiar example is colour. Any colour has both hue, saturation and brightness.

[2] But other curves, for instance circles, are not so simple. We need to cut them at two points. This is discussed again in Chapter 4.

another, and so on. We cannot add sounds, of course, but we can assign quantitative numbers to them indirectly by measuring related things: the length of a vibrating string, for instance. Pitch is a measurable quality, though its measures are derived from outside. The pitch continuum is one-dimensional. If we make middle C a boundary, this divides the continuum into two parts, and we cannot get from low C to high C without 'jumping' middle C.

However, sounds have another dimension, loudness. It will be helpful to ignore problems in ordering sounds by a system of relations so as to get a loudness continuum. Sounds considered in respect of pitch and loudness form a two-dimensional continuum. If we choose a boundary point here – it will be a sound of certain pitch and loudness – we will not divide the continuum for we may ascend the scale past this point at a different loudness or alternatively, pass through this level of loudness at a different pitch. Thus, we can pass round the chosen pitch–loudness point. We would have to remove all sounds of whatever loudness of middle C pitch to divide this continuum. Alternatively, we could take out all pitch ranges of a certain volume of noise. If loudness and pitch really fulfilled the strict conditions for systems of relations that weight does, then we could also define a pitch–loudness closed curve to divide the continuum. In fact, it is possible to whistle a closed sound curve. I leave that as an exercise for the reader.

Sounds differ from each other in timbre. This allows us to add a third dimension of sounds to the two we have already. Now it is again more than a little doubtful whether timbre, as a third dimension, can be ordered in a system of relations of the kind we need. But this is, again, a practical difficulty which we can overlook for the present purely illustrative purpose. To divide the continuum now we shall have to remove all the notes of a certain pitch over the whole range of timbre and loudness, or (if we could, but we cannot exactly) define a closed 'surface' which cuts the manifold. We have arrived at a rather complex system of relations, which is, fairly realistically, a three-dimensional 'space'.

6 What systems of relations really are

Now we have an explanation of what it means to speak of a system of relations. Systems of relations were seen to be sets possessing properties of order, measurability and dimension which exactly (or near enough) mirror properties of space itself. I hope the Leibnizian theory that space precisely is a system of relations has been made to seem attractive. There can be no doubt that its plausibility recommends it.

But what can we say for Leibniz's theory that systems of relations are ideal? It is quite unclear what that really means. We can make out an equally satisfactory result, however, if we can find ways in which the existence of the system depends on the existence of the things in it. Now, in fact, weight 'space' *is* constituted, in a clearcut sense, solely by things that have weight. Without the things, the field of the relation *being heavier than* would be the empty set. The weight 'space' was exactly the ordered set built from a non-empty field. There is no way to build up a systematic structure from the empty set. (This set allows a degenerate partial ordering, i.e. it has no structure.) Without things there can be no quality space.

In one sentence particularly, Strawson touches in a different way on the nerve of Leibniz's idea that systems depend on things. He writes about the spatial framework: 'The only objects which can constitute it are those which can *confer upon it its own fundamental characteristics*' (Strawson (1959), p. 39 [my italics]). It is quite obvious how this is true of systems of relations. The fundamental characteristics of its dimensions are plainly given to sound space by characteristics which belong first to the things ordered in the system. That is clear from the way the system was made up. Sound 'space' has dimensions of pitch, volume and timbre only because any particular sound has a pitch, volume and a timbre. The position is similar for measurement. No ordering relation, no operation of adding can be constructed without some characteristic of objects, their weight, pitch or whatever it may be. Any of these systems depends completely upon all the objects in it having, to some degree, the property (or properties) on which the system is based. The things do constitute the system in a perfectly definite way.

7 Space and sets

Now we have a quite detailed picture of what is meant by the phrase 'system of relations'. With the picture before us, we can see that it clearly does allow the sort of thing that Leibniz and Strawson, for example, want to say about it. It shows rather decisively that something both definite and correct is intended by the claims that a system of relations is ideal (abstract), dependent on the objects in it and constituted by them. So the negative claims which are the heart of the ontological reduction of space to things are correct as claims about systems of relations. But I now want to go on to argue that the analogy between true space and systems of relations is a very partial and faulty one. We have toiled hard at showing that the analogy is very close so far as it concerns how like they may be in structure. We will get much clearer about what space is if we see that in some respects it is utterly different from a system of relations, despite some similarities in others.

Before pressing on with these respects of difference, let us focus on one very distinct advantage which construing systems of relations as sets has. This is that there quite certainly exists a definite set theoretical object which the system of spatial relations is. Obviously, it is just the set of spatial objects as ordered by the various spatial relations (by their composition). The arguments that follow in §§7 and 8 are not intended to question whether there really is such an entity. They are aimed at showing that we cannot be talking about this object when we talk about space. With this in mind, let us turn to the question how this system, and any other system of relations, differs from space.

First, the way space contains the things in it is quite different from the way a set (a system) contains what is in it. Spatial things are not members of space;[1] they fill parts of space. No doubt the spatial relations among things in space define a complex set which is a system of relations; but we cannot say that space is this system. An easy way to understand this is to look at the way a Hasse diagram[2] is related to the set it is a diagram of. True space is like the diagram, not the set it portrays. The diagram works in this way: first it directly identifies a certain set – the set of circles in it. Let us call this unordered set the base set. Then the diagram also identifies an ordered set made up from pairs of members of the base set. The line connexions of the diagram show us which pairs this set contains. This last ordered set is the system of relations, and we might give further structure to it in ways we have seen already.

Now the diagram is certainly not this set. The diagram is a complex material thing, not an abstract set. It is also taken to be a diagram of some quite different ordered sets from the one it immediately identifies. For example, we might find a group of people ancestrally related in such a way that there is a 1–1 correspondence between people and circles of the diagram and in which the people are ancestrally related just as the circles are joined by downward lines. The diagram represents this set, not by

---

[1] But space is a set of points. This is discussed in detail in Chapter 8, §7.
[2] The reader may wish to refresh his memory of what what these diagrams are by a quick glance back at p. 11.

visual resemblances of any kind, but by an isomorphism between the ordered set which the diagram defines and the ordered set defined by the people and their kinships. If we look at space in the light of these ideas, we can say that objects in their spatial arrangement certainly define a base set (the set of material spatial things, obviously) and they also define an ordered set whose pair sets are fixed by how the things are arranged. But is it correct to say that space is this ordered set of objects? I think it is rather obvious that it is not correct to say so. Spatially arranged objects are something very like their own Hasse diagram, not the abstract set that this would be a diagram of. But we must look into this more carefully.

The Hasse diagram will turn out to be useful again, so we will do well to spend a little time on making clear how it is related to the sets it illustrates. The ordered sets are simply sets of ordered pairs (or triples etc.) of objects. Therefore, even the ordered set which the Hasse diagram is most directly linked with, which has its circles themselves as members, can be directly represented by other kinds of diagrams. We can change the Hasse conventions. Colour the very same circles with primary colours and order them according to how near the red end of the spectrum their colours place them. If circles *a* and *b* were connected by a downward line, we now make *a* redder than *b*. The new conventions identify the very same ordered set as before. It is surely clear that the Hasse diagram is not identical with the new diagram and that neither of them is identical with the set they identify. The Hasse diagram is really the sum, in the sense of the calculus of individuals, of the circles and their connecting lines. As I mentioned before, the diagram is a complex material thing. The lines and circles it contains are its parts not its members.

The point of all this about diagrams is to labour what is probably obvious: things are not together in space in the same way as they might be together in a set. Things are together in space in the immediate, concrete way in which they are together in the diagram. But the set is not like this. Its remoteness and abstractness emerge from the way different concrete objects can portray it. This obvious tactic exploits something that is not really at the heart of the issue: it depends on the ordered set giving only 'relations in extension'.[1] The core of it all is that space contains things in a quite different way from the way a set,

---

[1] We could deal with 'relations in intension' in a set theoretical system-of-relations way, too. If we did so, it would only be yet more obvious how remote all this is from physical space.

or a system of relations, does. Of course, it is not perfectly clear just what people meant by their use of the phrase 'system of relations'. It is not even certain that they had, themselves, something quite definite in mind, so that we can decisively exclude even the Hasse diagram from qualifying as the intended system of relations. But one thing is clear. It is the set theoretical explanation of the phrase, which we are now hotly pursuing, *that bears out the negative claims made about systems of relations.* That is about all that is clear.

8 Parts of space and members of sets

Even if this is obvious, it is still rather too intuitive. We can get something much firmer by turning back to the contrast between being a part and being a member. The circles in a Hasse diagram are parts of the diagram, not members of it. They are members of the ordered set, of the system of relations; they are not parts of it. Now objects in a spatial arrangement are certainly members of the set which we are calling the system of relations and which the arrangement defines. But objects are not members of space. On the other hand, they are not quite parts of space either.[1] But they do occupy regions which are parts of space. Every extended object fills some part of space. They do not fill or occupy parts of sets which are systems of relations. We have no clear sense to attach to the phrase 'fill part of a set' and it is far from clear how to construct one. Objects are not in space exactly as circles are in a Hasse diagram; but they are in space in a very similar way.

We can develop this point further by approaching it from another direction. Space is the sum of all regions. Only a few of these regions may be occupied by material things, of course. But if the objects are real and really extended then these occupied parts of space will be real too. So at least the sum of these real regions must itself be real and not ideal. Now, we can certainly say that space contains these parts as a set contains its subsets (indeed, we must say that). But this does not put us in anything like the position of saying that objects or regions are members of space. Space is not the set of its regions, but the sum of them. We seem far indeed from any analogy with systems of relations.

Is the Leibnizian point simply the observation that not all regions are occupied? That would not really help, as is clear from a little further

[1] Not according to the style of metaphysics which Leibniz needs, anyhow. But we cannot simply rule the idea out.

reflection on Hasse diagrams. The circles of a diagram are related across the connecting lines and the lines are an indispensable part of the whole complex material things. Similarly, objects are related by distance and direction relations across arcs, paths or intervals. A distance metric is a function which takes pairs of spatial points as its arguments and assigns a positive real number to each pair as its value. But this has geometrical interest only in that the points are regarded as the endpoints of some interval, path or arc across which the points are at a distance. (See Chapter 2, §7 on pathological adding and also Chapter 8, §3.) These intervals are parts of space, and the spatial relations which Leibniz, Strawson and others wish to use for reductive ends depend on there being such regions in or across which the relations are understood to hold.

There is another way of labouring the obvious. Since we are now dealing with a hardy perennial of the philosophical garden we had best be thorough, digging up every root, even if the task is somewhat tedious. Space contains all spatial objects. Thereby it contains all sums or heaps of spatial objects. For example, it contains all the body cells that make up you, and it thereby contains you. It is quite different with sets and systems of relations. The set of all your body cells does not contain you nor does it contain any part of you that is not a body cell. We can form the set that consists of your body cells together with you as a last separately entered member. But then it does not contain your foot or your hand.[1] However the space which contains your body cells does thereby contain every spatial part of you. Of course, we can fake up a set which contains all spatial things, all sums or heaps of them, and which is ordered by the composition of all distance and direction relations among these things. But it is not space because it does not contain the parts of things just by containing the things, but only by a separate and arbitrary stipulation as to membership.

Space is not a system of relations in the sense that looked so satisfactory before.

9 Doubts about quality spaces

It may strike the reader that our troubles with ordered sets are that sets are abstract and unintuitive, compared with the rather plausible and congenial ideas that first occur to us when we read Leibniz and Strawson.

[1] Your foot is not a subset of this set.

What happens if we give up a strict identification with sets and look for a less rigid and formal understanding of what systems of relations might be? First, we lose the important assurance that there is any definite entity, such as a set is, to fix our ideas upon. Next, it becomes less clear what it would mean to say that the objects constitute the framework, though it remains easy enough to see how they 'confer upon it its own fundamental characteristics' (Strawson (1959), p. 39). In some cases, at least, we might speak of a system of relations in the sense of a *quality space*. What this means gets clearer in the next paragraph. But before getting down to detail, I must point out that the arguments of this section are quite different in aim from those of §§7 and 8. There, I wanted to show that talk about the ordered sets is not talk about space. Here, I want to show that there is no such thing as a quality space of spatial relations.

Let us now look at some criticisms of systems of relations which do not depend on their set theoretical character. One of Strawson's relationist observations is this: 'The only objects which can constitute [the system or framework] are those which can *confer on it its own fundamental characteristics*' (Strawson (1959), p. 39 [my italics]). This is clearly enough a correct remark about systems of relations, as we agreed before. The three dimensions of the quality space of sound are pitch, volume and timbre. We can give these dimensions to the space only because any of the sounds we want to place in the system has a particular pitch, volume and timbre. These properties belong to the space only because they belong first to the objects it it. The objects really do bestow them on the space. Similarly, without some property such as weight or pitch we lack a basis for constructing ordering relations, operations of addition and so on. So the system rests on the fact that the objects that it orders have the properties on which it is based. So an object of weight w, for example, finds a place in the system which is marked out for it by its weight property, w.

We are interested in the suggestion that space proper is a constructed system of this kind. It is the range of objects that the system copes with and the uniqueness and elegance of its order and measure properties that are said to give true space its prominent place in our conceptual scheme. However, this idea overlooks a simple, but crucial, fact. Objects in space have a spatial position which has nothing to do with any quality whatever which they might possess, including spatial qualities. This is not true of any system of relations which makes up a quality space.

I think this is a decisive obstacle to regarding space proper as a quality space. We can set the obstacle up in other ways as well. An object may change its spatial position or its spatial relations to other objects without undergoing any qualitative change itself (or without a qualitative change in anything else). But an object may change its position in weight space, for example, only by changing its weight. (Or by changes in the weights of other things, perhaps.) Someone might now try to take spatial position as a quality of things just as weight is. But, unless we take spatial position to be some such quality as the perceptual state of a mentalistic monad, it seems perfectly clear that the spatial qualities of a thing are exhausted in its shape and size. Now if a thing changes its weight position and its weight relations, then it (or some collection of other things) must change its weight quality. But a thing in true space can change its position and its spatial relations without any change in its spatial qualities.

This may get a little clearer if we contrast space with something that really is a system of spatial relations that form a quality space. The set of cuboids (rectangular oblong prisms) can be ordered in a quality space with respect to their length, breadth and depth. The ordered set gives us a three-dimensional quality space of shape and size. Any cuboid will find its place in this system in virtue of its size and shape, not in virtue of its actual position in true space. It will change its place in the quality space if and only if it changes in size in one of the dimensions. That is quite unlike its position in space proper. It is not that there is no spatial quality space or system of spatial relations. It is that space is something quite different from these.

Now it is obvious which way we are going. We are really making the point that a quality space is a complex of universals: places in the system are places for qualities. But space proper is certainly not a complex of universals. Space is a particular, as Kant said it was. It is an individual thing of some kind. Several other contrasts between true space and quality spaces support the same conclusion.

First, any point in a quality space is a general point: any number of particular things might 'occupy' it. Two objects which have identical weight will occupy the same point in weight space, for example, and in sound space several tones which have the same pitch, volume and timbre will be at one position. True space does not merely happen to have only one thing at any place at a time, as a quality space may. Places in true space are not qualitative places, even though we allow that different things may fill the same place at different times. Since being at some place is not a matter of fulfilling general qualitative conditions there is no question whether qualitative conditions are met uniquely. That a thing has a unique place at a time certainly reflects the particularity of space. However, it is not the uniqueness but the particularity, the non-qualitativeness of space, that is my main point at present.

Secondly, in many-dimensioned quality spaces, the dimensions are usually qualitatively distinct, as pitch, volume and timbre are. They are different kinds of extension. But in space proper there are not different kinds of extension. If we put a coordinate system on the space we might speak loosely of the x, y and z coordinates as dimensions, just as we speak of length, breadth and depth. But these coordinates are merely different directions of extension; there is nothing which determines how we should choose them. We can freely rotate the axes; our system need not even be Cartesian. Strictly, the dimensions a space has is settled by how it may be disconnected by certain sorts of elements, as we saw before. Dimensions seen on the analogy of length, breadth and depth are not of the essence; dimensions that are different in kind are quite out of the question for true space.

However there are many-dimensioned quality spaces where this feature is much less clear than it is in sound space; for example, there is the quality space of differently sized rectangular prisms. We could hardly say that the dimensions there are simply qualitatively distinct. Still, dimensions there work like different qualities, even so. To avoid the embarrassment of several places for each thing we will need some arbitrary distinctions as to which edge of a cuboid is to count as its length. For example, we could choose its longest side as its length, its shortest side as its depth and the remaining side as its breadth. In that case the length dimension of the quality space *is* distinguished from the others by meeting a general condition – different points along it denote different extensions along a cuboid's longest side. Of course, it is not a qualitative general condition. But the dimensions are not at all like those of true space.

Thirdly, there can be homogeneous stretches of things in space, for example, a red line, a smooth surface, a volume of water at constant temperature and so on. A thing in space may be qualitatively similar throughout its extension. But there cannot be a weight line, or a sound area or volume unless we take this to be made up of a multitude of different things all closely similar (though there could never be enough of them if the space is continuous). If we think of the continuously changing weight of some object or a rising and loudening note we can get the analogue of a weight or sound line, though areas and volumes still remain a problem. The point is that extension in quality space calls for qualitative differences somewhere. Objects are simply extended in true space but not in quality space. In general, objects can only occupy points, not regions. The fate of variegated objects is worse, however, for they qualify for different points which are disconnected (unless the variegation is continuous). This problem about the occupation of quality space by things is quite different from the earlier one about the difference between being contained in a set and being contained in space.

It was Strawson's claim, if I understand it, that space and time are basic for our scheme of identifying particulars because the spatio-temporal framework of relations is marked out by its scope, its elegance and by the fact that each particular has a unique place in it. This is a mischaracterisation. Space and time are not marked out by advantages they possess as systems of relations. They are marked out by not being systems of relations at all, but rather particulars (better, spacetime is a particular). That is the importance of space and time in our scheme of concepts.

10 A non-qualitative relational space – the genealogical tree

We have left a large stone unturned. We have not thought about the idea Leibniz actually proposed: that space is like a genealogical tree.[1] Now a family tree is not a quality space since families are not formed by their members meeting some general qualitative conditions. People are in the same family because they are conjugally or ancestrally linked, not because they are like each other in some way. Clearly quality spaces allow us far more by way of structural analogies with space than we will ever be able to coax out of blood relationships. This was why we turned away from Leibniz's idea before. But we struck a formidable group of difficulties which might now seem to spring just from the qualitativeness of quality spaces. Perhaps we can escape them if we retreat to the genealogical tree even if we lose some structure by the way. But before we ask whether the difficulties follow us in this retreat let us try to find out what the analogy between space and family trees is supposed to offer us.

We talk about genealogical lines and so on, because a family tree is a

[1] See the quotation on p. 8.

Hasse diagram, quite literally. The tree is drawn on paper (on parchment for august personages). It represents the set of people in the family as ordered by conjugal and ancestral relations. The tree consists of inscriptions (proper names) rather than circles and there are downward lines across spaces on the chart which indicate relations of family descent. Even if we do not draw up actual diagrams our speaking of lines of descent springs directly from our envisaging just how we might write down the names and draw in the lines. As before, the tree diagram is a complex object which identifies a set (the set of inscriptions) which is the base of an ordered set which the diagram singles out by means of its own linear connexions. This last set is isomorphic with the ordered set of people in the family under the ancestral and conjugal relations. But for Leibniz this ordered set is itself an ideal construct out of the complex (scattered) object of the family as blood ties relate them. That is how the tree portrays the genealogical reality.

The more laborious set theoretical details of this account can hardly have been in Leibniz's mind when he likened space to the tree. But if this new suggestion is to avoid the problems of §§7 and 8, what can he have meant? Surely he must have had in mind at least these two facts: that the tree is just a representation of the system of family relations and that the representation is not carried by pictorial similarities between the family and the tree. However, we say of the family what is true only of the tree. In that case, the significant thing cannot be that there are no real lines in the offing; it is that actual lines in one object are being understood to represent what are not lines at all in another object (the sum, or scattered whole, of the people in the family). That is the sense in which genealogical lines and spaces are ideal. But how are we to understand this mistake as cropping up in the case of extended objects spatially related to one another? What can we make of the idea that space is merely a representation of extended objects spatially related?

If the Hasse diagram (the tree) is to represent the family, then it does so in being a real complex object playing a representing role. Just as a toy aeroplane is a real representative toy, though not really an aeroplane, so a Hasse diagram is a real spatial diagram, though not really a family. If the analogy is to hold, then Leibniz ought to regard space as a real something or other playing a representing role. But a real what? Space could only provide us with a properly spatial representation of extended things in a spatial arrangement. But this is not a representation; it is the thing itself. Space is its own Hasse diagram; this is a degenerate form of representation, at best.[1] It is pretty thoroughly obscure how we should try to grasp the analogy Leibniz has offered us.

But even if we were quite clear about how to understand the suggestion Leibniz made, we would be no better off than we were with quality spaces. We get the very same problems as before. There is no way to make sense of a thing's being extended in genealogical space, of filling up a region in the family system. Since we have so little structure provided by the system of blood ties perhaps we ought not to say that people occupy only points in it. We cannot make out a distinction between points and regions yet. But it is pretty obvious that any device which would yield up continuity from the 'space' of family relations would confine people to points in it. Of course, the downward ordering of names in the tree need not be based just on genetic causal relations; we also use temporal ones. We might be inclined to allot a region of the tree to a name as a function of the person's lifespan. This would obscure the fact that the family relations themselves do not provide for extension. Leibniz cannot have intended to trade on temporal extension, however, since time is just as problematic as space, nor is there any hint at all that he wants us to bring temporal extension into consideration: he suggests only that we measure the space by counting the generations as causal episodes. This falls very far short of what we need to explain quantitative measures in space, however.

Genealogical trees are not quality spaces. But they do have a generality of positions marked out in them, just as quality spaces do. Several individuals can occupy the same place in the hierarchy. Suppose a couple produce several offspring who die without marriage or issue. The genealogical relations cannot distinguish them, though they can be separated by priority of birth (or death), by being in different places at the same time, and so forth. But these ways of sorting them out are of no use to us. In the system proper, they fill the same place. This dual occupancy does not depend on the accidents that the siblings die single or childless. Suppose the several indiscernible male offspring of certain parents marry the indiscernible equally numerous female offspring of other parents, each pair producing offspring. We can distinguish various places in genealogical space, but we cannot separate the people who fill them. Obviously this could ramify. Of course dual occupancy here is accidental, improbable and precarious. Let one indiscernible descendant marry a well distinguished individual and a whole subsystem can

---

[1] We could use one space to represent another, of course. But this can hardly be envisaged as a metaphysical reduction.

thereby be marked out. Nevertheless, positions in genealogical space are general, if not qualitative, whereas positions in true space are particular.

At this point we might be tempted to wonder how secure the principle is that spatio-temporal relations place things uniquely. Certainly waves may cross one another and interfere; there seems no reason to disallow their being in some place at one time.[1] But we can deal with waves in a material medium, such as water, by regarding the water (or its molecules) as the thing and the wave as a shape the water has momentarily in a certain region. It is not nearly so clear that we can deal as neatly with the interferences and overlaps of light waves or the waves associated with fundamental particles. These are states of a field, and while a field is an entity for physics it is not simply a sea of material stuff. It may be that we can give a field theory account of matter and so of material objects. In this story, the universe is one indivisible thing, a field or plenum, in which we can locate evolving states. This might well undermine the principle that only one 'thing' can be in a place at a time. But this angle on the problem cannot be exploited in the present context. For here the basic metaphysical assumption is that things are ontologically first and fields and space are to be analysed in terms of them. Leibnizians cannot afford to abandon the spatio-temporal uniqueness of things at this point.

Leibniz also suggested a very ingenious model for motion in family space. The idea was metempsychosis: it is souls that migrate, not people.[2] But this is not strictly analogous to the motion of things through space. The entities whose positions in genealogical space are fixed by family relations are not the things that 'move', whereas in space the things that move are just the things that are spatially related to one another. Leibniz's model is a dualistic one. The soul is attached to one or another body in some inscrutable way. Each body is generated by the physical copulation of parents; it then generates its own offspring by further bodily coupling with another. Finally it dies, which completes the range of things to which it is related by direct parental or conjugal ties. The soul is thereby released for fresh attachment to a new body whose different blood relationships define a different place for it in the system. Blood relationships attach in a secondary way to the soul only because it is linked by some quite different tie to a body. The body cannot move in family space. Only the inscrutable body–soul tie allows for the motion of something. It is hard to see how this really helps the problem of explaining motion by way of systems of relations.

[1] I am indebted to C. A. Hooker in this paragraph.   [2] See p. 8 above.

Family space, then, offers us rather less than either sets or quality spaces do. We lose a great deal of the structure that sets or quality spaces provide us with. We cannot construct a continuous space, we have no foothold for many dimensions and there is no model even for addition in sight. In the end, therefore, it seems that systems of relations have mainly a negative interest: they make it clear in what way they are crucially unlike space. For it is not that space is continuous, measurable or many-dimensioned that is the core of the matter. It is that space is particular and quality space (like the less interesting genealogical space) is general. It is also that space is somehow concrete (things are *in* it). But sets are abstract.

The moral to be drawn is that space is a particular entity, then. But perhaps it is easier to draw it than to understand it. We will continue to feel uneasy about having space as a thing in our ontology so long as the idea of its particularity or concreteness remains vague or confused. We know some of the attributes which space may have: we take it to be continuous, quantitative and many-dimensioned. But general things can have these attributes too – quality spaces can, as we saw. Somehow we need to find some attributes of space that seem to be the attributes only of particulars. We need to describe space so that it gets clear that it really is an individual and, even, a concrete thing. A space has a shape. That sounds concrete and particular enough, but it cries out for explanation on its own account. The next three chapters aim to provide the main outlines of this. Our first clue comes from an apparently humdrum property of the most familiar of things, our hands. Let us look into the foundations of the distinction between left and right.

2 Hands, knees and absolute space

1 Counterparts and enantiomorphs

My left hand is profoundly like but also profoundly unlike my right hand. There are some trifling differences between them, of course, but let us forget these. Suppose my left hand is an exact mirror-image replica of my right. The idea of reflection deftly captures how very much alike they might be, while retaining their profound difference. We can make this difference graphic by reminding ourselves that we cannot fit left gloves on right hands. This makes the point that one hand can never occupy the same spatial region as the other fills exactly, though its reflection can. Two objects, so much alike yet so different, are called 'incongruent counterparts'.

In my usage that phrase expresses a relation among objects, just as the word 'twin' does. Thus, no one is a twin unless there is (or was) someone to whom he is related in a certain way. Call this relation 'being born in the same birth'. Then a man is (and has) a twin if and only if he is born in the same birth as another. Let us call the relation between a thing and its incongruent counterpart 'being a reflected replica'. Then, again, a thing *is* an incongruent counterpart if and only if it *has* one. My right hand has my left hand, and the left hands of others, as its incongruent counterparts. If people were all one-armed and everyone's hand a congruent counterpart of every other, then my hand would not be (and would not have) an incongruent counterpart.

But incongruent counterparthood gets at something further and deeper than twinhood does. No actual property belongs to me because I could have been born in the same birth as another. But all right hands do share a property, just because incongruent counterparts of them are possible. Though there are incongruent counterparts of my left hand, it is not their existence that makes it left. It appears to be enough that they could exist for my hand to have this property. Let us express this new further and deeper idea by calling hands 'enantiomorphs' and by saying that they have handedness. Then that is not an idea that depends on relations of hands to other things in space, as incongruent counter-parthood does.

We can show this in various ways. A well-worn method is to invent a possible world, say, one where all hands are right, as I did a moment ago. A more vivid possible world is one that contains a single hand and nothing else whatever. This solitary hand must be determinate as to being either left or right. It could not be indeterminate, else it would not be determinate on which wrist of a human body it would fit correctly, that is, so that the thumb points upward when the palm touches against the chest. Enantiomorphism, then, is not a relation between objects that are in space.

Let us not make too much mystery out of this. We can say how hands, or anything else, come to pose a problem: they have no centre, axis or plane of symmetry. The sphere has all three: many common shapes have at least a plane of symmetry, the human body being an example. But the failure of symmetry yields no difference between left and right hands, since each is asymmetric. It gives just a necessary condition of enantiomorphism.

2 Kant's pre-critical argument

These ideas originate in Kant, of course. Though enantiomorphs crop up in the *Prolegomena*, they feature in and around the *Critique* as illustrating that space is the form of outer sense. But in an early paper (Kant, 1968) of 1768, 'Concerning the Ultimate Foundations of the Differentiation of Regions in Space', he used them to argue for the ontological conclusion that there is absolute space. It is this earlier argument that this chapter is all about. Kant rightly regarded his paper as a pioneering essay in *Analysis situs*, or topology, though an essay equally motivated by metaphysical interests in the nature of space. He recognised only Leibniz and Euler as his predecessors in this geometrical field. The argument has found little favour among geometers and philosophers since Kant first produced it. I guess no one thought he was even half-right as to what enantiomorphism reveals about space. Kant himself had second and even third thoughts about his argument. My aim is to show that his first ideas were almost entirely correct about the whole of the issue.

This is not to deny that Kant's presentation is open to criticism. He was ignorant of several important and relevant facts which emerged later.[1] But knowledge of these leads me, at least, only to revise the detail

[1] A clear and lively account of many of the facts can be found in Gardner (1964).

of his argument, not to abandon its main structure and content. In the remainder of this section I will set out a version of Kant's argument. I try to follow what I think were his intentions closely and clearly, but in such a way as to indicate where further comment and exposition are needed. Hence this first statement of the argument is rather bald.

A_1: Any hand must fit on one wrist of a handless human body, but cannot fit on both.

A_2: Even if a hand were the only thing in existence it would be either left or right (from A_1).

A_3: Any hand must be either left or right (enantiomorphic) (from A_2).

B_1: Left hand and right hand are reflections of each other.

B_2: All intrinsic properties are preserved under reflection.

B_3: Leftness and rightness are not intrinsic properties of hands (from B_1 and B_2).

B_4: All internal relational properties (of distance and angle among parts of the hand) are preserved in reflection.

B_5: Leftness and rightness are not internal properties of hands (from B_1 and B_4).

C_1: A hand retains its handedness however it is moved.

C_2: Leftness and rightness are not external relations of a hand to parts of space (from C_1).

D: If a thing has a non-intrinsic character, then it has it because of a relation it stands in to an entity in respect of some property of the entity (from A_3, B_3, B_5, C_2, D).

E: The hand is left or right because of its relation to space in respect of some property of space.

Kant places several glosses on E. He claims that it is a hand's connexion 'purely with absolute and original space' (Kant (1968), p. 43) that is at issue. That is, space cannot be The Void, a nonentity, since only something existing with a nature of its own can bestow a property on the hand. Again, to have made a solitary right hand rather than a solitary left would have required 'a different action of the creative cause' (p. 42), relating the hand differently 'to space in general as a unity, of which each extension must be regarded as a part' (p. 37).

The premise D is not explicit in Kant. It is, quite obviously, a highly suspect and rather obscure claim. But I think Kant needed something of this very general nature, for he did not really know what it was about space as a unity that worked the trick of enantiomorphism. It will be possible to avoid this metaphysical quicksand if we can arrive at some clearer account of just what explains this intriguing feature.

3 Hands and bodies: relations among objects

A number of relationists have replied to Kant's argument by claiming
flatly that a solitary hand must be indeterminate as to its handedness.
But it is very far from clear how a hand could possibly be neither left
nor right. The flat claim just begs the question against Kant's lemma A,
which argues that a hand could not be indeterminate as to which wrist
of a human body it would fit correctly. Nevertheless, there is an influential argument to the effect that this lemma is itself a blunder. It is
instructive to see what is wrong with this relationist contention.

Kant's reason for claiming that it cannot be indeterminate whether
the lone hand is right is perhaps maladroit because it suggests that the
determinacy is constituted by a counterfactual relation to a human
body. The suggestion is hardly consistent with Kant's view that it is
constituted by a relation between the hand and space. However, it
provides Remnant[1] (1963) with a pretext for foisting on Kant an essentially relationist view of the matter. He takes Kant to be offering the
human body as a recipe for telling whether the lone hand was left or
right. He shows quite convincingly that even the introduction of an
actual body into space will not let us tell whether the hand is left or
right. For suppose a handless body is introduced. The hand will fit on
only one of the two wrists correctly, i.e., so that when the arm is thrown
across the chest with the palm touching it, the thumb points upward. But
this does not settle whether the hand is left or right unless we can also
tell whether it is on the left or right wrist that the hand fits. But we can
be in no better position to settle this than to settle the original question
about the single hand. Possibly we are in a worse position because the
human body (handless or not) is not enantiomorphic (except internally,
with respect to heart position, etc.). Even if every normal human body
had a green right arm and a red left (so that it became enantiomorphic
after a fashion), that would not help. For incongruent counterparts of it,
with red right and green left arms are possible. News that the hand fits
the green wrist enables us to settle nothing about its rightness unless we
know whether the body is normal. Settling this, however, is just settling
handedness for a different sort of object. Thus, it is concluded, a solitary
hand is quite indeterminate in respect of handedness.

What Remnant shows, in these arguments, is that no description in
terms of relations among material things or their material parts ever
distinguishes the handedness of an enantiomorph. This is an objection

[1] The paper is cited as definitive by Gardner (1964) and Bennett (1970).

to Kant only if his maladroit reason is construed as a lapse into the view that what makes the hand left is its relationship to a body. Since this casts his reason as a contrary of the conclusion he draws from it, the construal is improbable. Remnant takes Kant to have been guilty of the blunder of supposing that, though it is impossible to tell, of the single hand, whether it is left or right, it is quite possible to tell, of a lone body, which wrist is left. It would have been a crass blunder indeed. But Kant never says that we can tell any of these things. In fact, he denies it. Insofar as Remnant does show that relations among the parts of the hand and the body leave its handedness unsettled, he confirms Kant's view. His attack on Kant succeeds only against an implausible perversion of the actual argument.

Kant's introduction of the body is aimed, not at showing the hand to be a right hand or a left hand, but at showing that it is an enantiomorph. It shows this perfectly clearly. For the hand would certainly fit on one of the wrists correctly. It seems equally certain that it could not fit correctly on both wrists. Whether it fits a left wrist or a right is beside any point the illustration aims to make. Though this expository device may suggest a relationist view of enantiomorphism, it does not entail it. It is simply a graphic, but avoidable way of making the hand's enantiomorphism clear to us. The idea that a hand cannot be moved into the space that its reflection would occupy is certainly effective, too. But it is less striking and less easily understood (I shall say more about it later). Kant was quite well aware that this method was also available and, indeed, used a form of it in his 1768 paper.

Thus the objection to the determinacy of handedness in solitary objects fails.

4 Hands and parts of space

John Earman (1971) charges Kant with incoherence. No relation of a hand to space can settle handedness. Kant appeals to a court that is incompetent to decide his case. It is useful to look into Earman's objection.

As I understand Earman, he argues as follows: we can plausibly exhaust all relevant spatial relations for hands under the headings of internal and external relations. Kant takes the internal relations to be solely those of distance and angle which hold among material parts of the hand. But these do not fix handedness, since they are invariant under reflection, but handedness is not (B_5). What makes Kant's argument incoherent is that the external relations (to the containing space) cannot fix it either. For the external relations can only be those of position and orientation to points, lines, etc., of space outside the hand. (Incidentally, whether these parts of the container space are materially filled or not appears to make no relevant difference.) But for every external relation a right hand has to a part of space, a left hand has just that relation to a similar part, as reflection makes quite clear. In any case, changes in these relations occur only if the hand moves through the space. However, although movement through the space can alter these relations, it cannot alter handedness (C_2). Evidently none of these relations settles the problem. This presents us with an unwelcome parody of Kant's argument: since handedness is not settled by either internal or external relations in space, it must consist in the hand's relation to some further well-defined entity beyond space (Earman (1971), p. 7).

That is an unlucky conclusion. We will have to think again.

Earman offers us a way out of the wood. The spatial relations we have been looking at do not exhaust all that are available. He suggests that we say, quite simply, that *being in a right configuration* is a primitive internal relation among parts of the hand. If we could say that it would certainly solve the problem in a very direct fashion. It would be a disappointing solution since we cannot explain the difference between left and right by appeal to a primitive relation. But, in fact, there is no such relation, as I will try to show in §§6 and 7.

However, my interest in the present section lies in other things that Earman has to say. He frankly concedes that Kant was well-aware of the kind of objection (as against the kind of solution) that he offers, so that, in this respect, his criticism is 'grossly unfair'. For Kant did insist that it is 'to space in general as a unity' that his argument appeals. Earman says (p. 8) that he does not see how this helps. But that merely invites us to take a closer look.

We can get some grip on the idea of 'space in general as a unity' by taking a quick trip into geometry. We need not make heavy weather of the rigour of our journey. A rigid motion[1] of a hand is a mapping of the space it fills which is some combination of translations and rotations. Such mappings make up an important part of metrical geometry.

---

[1] Naively, a rigid motion is a movement of a thing which neither bends nor stretches it. As a mapping, it is just a function which takes us from the space of the thing to a space into which it might be rigidly moved. Reflection is a mapping but it is not a motion.

A reflection is also a mapping of a space, and, like a rigid motion, it preserves metrical features. This last fact, so important for Kant, is easily seen by supposing any system of Cartesian coordinate axes, x, y and z. A reflection maps by changing just the sign of the x (or the y or the z) coordinate of each material point of the hand. In short, it uses the y–z (or the x–z or the x–y) plane as a mirror. Thus it preserves all relations of distance and angle of points in the hand to each other, since only change of sign is involved. Thus the lemma B_1, B_4, B_5 is sound.

Now we can express the idea of enantiomorphism in a new way which has nothing to do with possible worlds or with the relation of one hand to another (actual or possible) hand or body. It does, however, *quantify over all mappings* of certain sorts. We can assert the following: Each reflective mapping of a hand differs in its outcome from every rigid motion of it. That is a matter of space in general and as a unity. (Thus lemma C can be gained virtually by definition.) This quantification over the mappings seems to have nothing to do with any object in the space; not even, really, with the hand that defines the space to be mapped. Though this terminology is much more recent than Kant's, the ideas are old enough. I see no reason to doubt that they are just what he intended. Space in general as a unity is exactly what is at issue.

5 Knees and space: enantiomorphism and topology

Kant was right. The enantiomorphism of a hand consists in a relation between it and its containing space considered as a unity. This is more easily understood if we drop down a dimension to look at the problem for surfaces and figures contained in them. We need to see how hands could fail to be enantiomorphs.

Imagine counterpart angled shapes cut out of paper. They are like but not the same as L's, since L's are directed. Let me call them knees, for short (but also, of course, for the legitimacy of my title). They lie on a large table. As I look down on two of them, the thick bar of each knee points away from me, but the thin bar of one points to my left, the thin bar of the other to my right. Though the knees are counterparts, it is obvious that no rigid motion of the first knee, which confines it to the table's surface, can map it into its counterpart, the second knee. Clearly, this is independent of the size of the table. The counterparts are incongruent. The first knee I dub left. The second is then a right knee.

Their being enantiomorphs clearly depends on confining the rigid

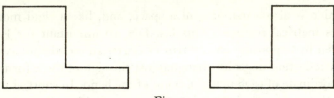

Fig. 4

motions to the space of the table top, or the Euclidean plane. Lift a
knee up and turn it over, through a rigid motion in three-space, and
it returns to the table as a congruent counterpart of its mate. That the
knees were incongruent depended on how they were put on the table or
how they were in the space to which we confined their rigid motions.

A different picture is more revealing. Suppose there is a thin vertical
glass sheet in which the knees are luminous angular colour patches that
move rigidly about. They are *in* the sheet not *on* it. Seen from one side
of the sheet, a knee will be, say, a left knee. But move to the other side
of the sheet and it will be a right knee. That is, although the knees are
indeed enantiomorphs in being confined to a plane of rigid motion, any
knee is nevertheless quite indeterminate as to being a left rather than a
right knee, even granted the restrictions on its motion. It could hardly
be clearer, then, that nothing intrinsic to an object makes it left or right,
even if it is an enantiomorph. Our orientation in a higher dimensional
space toward some side of the manifold to which we have confined the
knee prompts our inclination to call it left, in this case. It is an entirely
fortuitous piece of dubbing. Hence, if the knees were in a surface of just
one side (and thus in a *non-orientable manifold*) they would cease to be
enantiomorphic even though confined to rigid motions in that surface.
(Thus B_3 is correct independently of B_1 and B_2 and despite Earman's
objection.)

There is a familiar two-dimensional surface with only one side: the
Moebius strip (see fig. 5). If we now consider knees embedded in the
strip, they are never enantiomorphic. A rigid motion round the circuit
of the strip (which is twisted with respect to the three-dimensional
containing space) turns the knee over, even though it never leaves the
surface. However, the Moebius strip might be considered anomalous as
a space. It is bounded by an edge. We need the space of Klein's bottle,
a closed finite continuous two-space (see fig. 6).[1] Rigid motion of a knee

[1] This surface always intersects itself when modelled in Euclidean three-space.
But this does not rule it out as a properly self-subsistent two-space. It can be
properly modelled in four-space, for example.

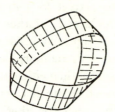

Fig. 5. Moebius strip.   Fig. 6. Klein bottle.

round the whole space of Klein's bottle maps it onto its reflection in the space. Knees are not enantiomorphic in this one-sided, non-orientable manifold. They are indifferent as to left or right. Let us say that here they are *homomorphic*.

These general results for knees as two-dimensional things have parallels for hands as three-dimensional things. It seems to be pretty clearly the case that, as a matter of fact, there is no fourth spatial dimension that could be used to turn hands 'over' so that they become homomorphs. No evidence known to me suggests that actual space is a non-orientable manifold. But it cannot be claimed beyond all conceivable question that hands are enantiomorphic, and it is not too hard a lesson to learn how they could be homomorphic. So Kant's conclusion A_3, a principal lemma, is false. But this has nothing to do with whether there is one hand or many. It has nothing to do with an obscure indeterminacy that overtakes a hand if there are no other material things about. It is false because spaces are more various than Kant thought.

Thus, whether a hand or a knee is enantiomorphic or homomorphic depends on the nature of the space it is in. In particular, it depends on the dimensionality or the orientability, but in any case on some aspect of the overall connectedness or topology of the space. Whether the thing is left depends also on how it is entered in the space and on how the convention for what is to be left has been fixed. Kant certainly did not see all this. Nevertheless, it should be obvious now how penetrating his insight was.

6 *A deeper premise: objects are spatial*

Clearly enough, Kant's claim that a hand must be either left or right springs from his assumption that space must have Euclidean topology, being infinite and three-dimensional. The assumption is false, and so is the claim. This suggests an advantageous retreat to a more general disjunctive premise for the argument, to replace A_3. Rather than insist that the hand be determinately either left or right, we insist rather that it be determinately either enantiomorphic or homomorphic. Thus, if there were a handless human body in the space, then either there would be a rigid motion mapping the hand correctly onto one wrist but no rigid motion mapping it correctly onto the other; or, there would be rigid motions some of which map it correctly onto one wrist and others which correctly map it onto the other wrist. Which of these new determinate characters the hand bears depends, still, on the nature of the space it inhabits, not on other objects. The nature of this space, whether it is orientable, how many dimensions it has, is absolute and primitive.

What underlies this revision of Kant's lemma A is the following train of ideas. We can dream up a world in which there is a body of water, without needing to dream up a vessel to contain it. But we can never dream up a hand without the space in which it is extended and in which its parts are related. To describe a thing as a hand is to describe it as a spatial object. We saw the range of spaces a knee might inhabit to be wide; the same goes for spaces in which a hand might find itself. So dreaming up a hand does not determine which space accompanies it, though Kant thought it did. But it does not follow that there could be a hand in a space that is indeterminate (with respect to its global connectivity, for example). We can describe a hand, leaving it indeterminate (unspecified) whether it is white or black. But there could not be a hand indeterminate in respect of visual properties. Like air in a jar, even an invisible hand can be seen to be invisible, so long as we know where to look. (I am here shuffling under a prod, itself invisible here, from David Armstrong.) No considerations mentioned yet admit of a hand that could be neither enantio- nor homo-morphic. There seems no glimmer of sense to that expression.

But I spoke just now of there being a wide range of spaces that hands or knees might occupy. What does this mean, and does it offer a route to the relationist between the alternatives I am pushing? What it means is that we can describe a knee, for example, as a mass of paper molecules (or continuous paper stuff) which is extended in a certain metrical two-space. We can regard this two-space as bounded by extremal elements that make up edges (or surfaces for the three-space of a hand). These elements define the shape the mass of matter is extended in by limiting it. A wide variety of global spaces can embed subspaces isomorphic (perhaps dilated) with our knee- and handspaces. In short, a kneespace is a bit of our ordinary space. Nothing mooted here is meant to suggest that a hand might somehow be taken from one space to another while being spaceless in the interim. Let us, for a moment, toy with the idea that a kneespace need not be a subspace at all, but that it could just end at its material extremities without benefit of a further containing space. Does the hand or knee become indeterminate as to enantiomorphism if we consider it just in its own handspace or kneespace?

Kant evidently feared that it would, and it might seem that he was right. Can't we argue that, for the hand or knee to be either enantio-morphic or homomorphic, there must be enough unified space to permit both the reflection and some class of rigid motions to be defined in it? Otherwise the question whether any rigid motion maps the hand onto its reflection does not have the right kind of answer to yield either result. But this is wrong. What counts is not whether the particular object has a reflection or whether it can be rigidly moved in the kneespace, but whether suitable objects in general do. It depends, first, on the space, not on the object. Both handspaces and kneespaces are orientable spaces. This is easy to see by imagining a much smaller hand or knee in the space and considering its reflections and the class of its rigid motions in that limited space. Clearly a hand in a handspace is enantiomorphic. This does not mean that hands are intrinsically enantiomorphic. It means that handspaces are certain kinds of spaces. Any hand must always lie in a handspace at least as a subspace, but there is no necessity about which space it is a subset of. Nor does there seem to be anything to say about the hand as material which could determine even so small a thing as whether the space extends beyond the matter of the hand or is confined to it. Dreaming up a hand means dreaming up a space to contain it, of which the being and nature are independent and primitive. It is introduced in its own right as a well-defined topological entity.

An oddity here is that, though a hand filling its handspace can be a dilated incongruent counterpart of a small hand in its space, we could never compare or contrast one hand in its own handspace with another in its own handspace. That requires mutual embedding in a common space. I conclude that leftness is not a primitive relation with respect to which hand parts are configured. I touch on this topic again in §7.

But handspaces or kneespaces as anything other than subspaces have, so far, been mere toys of our imagination. The thought that space could just come to an end is one at which the mind rebels. Let us express our distaste for such spaces by calling them pathological.[1] I want to do more now than toy with the idea of pathological spaces. It was an important conviction in Kant's mind, I think, that pathological spaces cannot be the complete spaces of possible worlds. Though I can think of no strong defence for this deep-lying conviction, which most of us share, I can think of an interesting one. At least, it interests me. Why might one think that space cannot have boundaries?

The thought that space cannot simply come to an end is ancient. The argument was that if, though impossible, you did come to the end you could cast your spear yet further. This challenge is quite ineffective against the hypothesis that there is just nowhere for the spear to go. But what the challenge does capture, rather adroitly, is the fact that we cannot envisage any kind of mechanics for a world at the point at which moving objects just run out of places to go. What would it be like to push or throw such an object? We can't envisage. It would be unlucky if this boggling of the mind tempted us to regard pathological spaces as contradictory. To go Kant's way on this is to go the way of synthetic necessary truth. In the 1970s, the prospects for following that road are dim. But there are prospects, more tangible and, perhaps, a bit brighter, since Kripke's semantics for modal logic. It is tempting to pursue this defence of Kant's inference from pathological spaces to global ones which properly contain them. It would be a protracted and rather tangential undertaking, however. The best defence for the disjunction enantiomorphic/homomorphic is to argue, as I did earlier, that hands and knees must be either homomorphic or enantiomorphic whether their containing spaces are pathological or not. This rests the disjunction on the deeper premise that hands are spatial objects and there can be no hand without a space in which it is extended.

Nevertheless, I am inclined to offer the suggestion that, when it is our

[1] More technically, I believe that we find a space pathological when it can be deformed over itself to a point, or to a space of lower dimension. Topologists call such spaces contractible. (See Patterson (1956), p. 74). The sphere as a three-space is deformable through its own volume to a point, but the surface of the sphere, as a two-space, cannot be deformed over its own area to a point; so it is not pathological or contractible. The Moebius strip can be deformed across its own width to a closed curve (of lower dimension). It was for this reason that I moved, earlier, to Klein's bottle as a non-contractible proper space.

task to conceive how things might be, as a whole, we should ask for what I will call an *unbounded-mobility* mechanics. (I intend the phrase to recall Helmholtz's 'free mobility', which he used to express the possibilities of motion in spaces of constant curvature.) That would rule out pathological spaces for possible worlds.

7 Different actions of the creative cause

So far I have not let the relationist get a word in edgeways. But his general strategy for undermining our argument is pretty obvious. He must claim that there cannot be a space that is a definite topological entity unless there are objects that define and constitute it. Of course, he has to do more than simply to assert his claim; he has to make it stick. That needs at least two things. First, he needs to show how bringing objects into the picture can settle topological features of a space in some way – for example, settle the feature that it is an orientable manifold. Then he has to show, next, why only objects can give it. The second of these tasks has the virtue of familiarity, at least, but I know of no discussion whatever of the first.

Let us pick, for our example, the case that still lies before us, of distinguishing orientable from non-orientable manifolds. Suppose we begin with hands or knees in their pathological or limited spaces. These are orientable spaces which can be subsets of non-orientable spaces. How could relations among these hands or knees make some wider space orientable? Knees are simplest. I will talk about them.

Suppose there are two knees, each in its own pathological space, neither being primitively taken as part of a wider, mutually inclusive manifold. The supposition is expressed more accurately, perhaps, in this sentence:

$(\exists x)(\exists y)(x$ is a knee in a kneespace & y is a knee in a kneespace & $x \neq y)$

According to this hypothesis there is no primitive spatial relation between the knees. Let us suppose further that some change in one knee causes a change in the other. If cause is the transmission of an effect, then this suggestion of a causal connexion presupposes that there is, after all, a more primitive spatial relation between the things. But if cause is not the transmission of an effect, then how does the supposition of a causal connexion even begin to relate the things in some derivative spatial way? I see no glimmer of light in this sort of approach. We can add knees to

knees in this fashion till the cows come home and be no nearer constituting a wider non-orientable manifold, or indeed any wider manifold
at all. The trouble is that it is not just our kneespaces that are pathological; so is the method of adding just described. What seems always
to have been taken for granted is that adding objects is putting them
together in the same unified space. I do not see how that can fail to mean
that we posit paths (continua of points) across which they are related
(e.g. across which they are at some distance). The reason for always
taking that for granted, I guess, is the belief, shared with Kant, that
pathological spaces are not possible.

Consider Melissus' paradox: By virtue of there being nothing between
two distinct objects, there is empty space between them; that is, there is
something between them. The remarks just made about pathological
adding and pathological spaces should make it clear that I rest no weight
on this intriguing argument. Nevertheless, I do claim that in all possible
worlds, as we ordinarily envisage them in philosophy, there always is a
path between two objects that do not touch each other.

Pathological adding cannot help the relationist. Let us press on with
standard adding. A model will help us focus our ideas. Suppose there are
two strips of paper, one white and the other red. We cut several knees
out of the red paper and we intend to embed these in the white strip by
cutting appropriate shapes out of the white paper and fitting the knees
into them. Clearly, once we have made a cut in the white strip to admit
the thick bar of a knee, there are two directions in which we can cut to
admit the thin bar.

Well then, a space is orientable if it can[1] be covered by an array of
directed entities in such a way that all neighbouring entities are like-
directed. In our case, the question is whether we can make the white
strip orientable or non-orientable by entering the knees in some ways
rather than in others. The answer is that these entries have no bearing
on the matter at all. Suppose our white strip is joined at its ends to form
a paper cylinder. Then we can cover it with knees in such a way as to
illustrate its orientability. But clearly this illustrates without constituting
its orientability. That is obvious once we see that we can cut across the
strip, give it a twist and rejoin it without changing the way any knee is
embedded in it. What we now have is a Moebius strip which is non-
orientable. Some pair of neighbouring knees will be oppositely directed.
That is solely a matter of how the space (the white strip) is pathwise

[1] Notice that a space's *non*-orientability is never constituted by how it *is*
covered.

connected globally. It is quite irrelevant how many knees are embedded in the strip and how their shapes have been cut out for them.

This makes it look, more than ever, as if the space as a definite topological entity can only be a primitive absolute entity; that its nature bestows a character of homomorphism, leftness or whatever it might be, on suitable objects. My conviction of the profundity of Kant's argument rests on my being quite unable to see what the relationist can urge against this, except further relationist dogma. As always, of course, that might mean just that I still have lessons to learn.

But so do relationists. The difficulty of our going to school with open minds on the matter is strikingly shown in Jonathan Bennett's paper 'The Difference between Right and Left' (1970). His paper is devoted largely to the question whether and how an English speaker whose grasp of the language was perfect, save for interchange of the words 'left' and 'right', could discover his mistake. He has to learn it, not ostensively, but from various descriptions in general terms relating objects to objects. Bennett concludes that the speaker could not discover his error that way, though he could discover a similar error in the use of other spatial words, such as 'between'. It is a long, careful, ingenious discussion. But it is an utterly pointless one. Bennett states on p. 178 and again on p. 180 that in certain possible spaces (some of which we know) there may not be a difference between left and right. So how could one possibly discover 'the difference' between left and right in terms of sentences that must leave it entirely open whether there is any difference to be discovered? That cannot be settled short of some statement about the over-all connectivity of the space in which the things live. The same may be said, moreover, of Bennett's discussion of the case of 'between'. For, given geodesical paths on a sphere (and Bennett is discussing air trips on earth more or less), the examples he cites do not yield the results he wants. The familiarity of relationist approaches appears to fixate him and prevent the imaginative leap to grasping the relevance of those known global spatial results which clarify the issue so completely. No doubt relationism has some familiar hard-headed advantages. But it can also blinker the imagination of ingenious men. It can make them persist, despite better knowledge, in digging over ground that can yield no treasure. Empiricist prejudice is prejudice.

Let us get back to my paper strip and the knees embedded in it. It is a model, of course. It models space by appeal to an object, one which also defines what is essential to the space. That is, the subspaces, paths, and mappings that constitute the space are modelled by the freedoms and limitations provided by the constraint of keeping objects in contact with the surface of the modelling body (or fitted into it). This modelling body is embedded in a wider true space, which enables the visual imagination to grasp the space as a whole. Realists do treat space just like a physical object in the sense of such analogies. A space is just the union of path-wise connected regions.

The model also shows us a distinction already mentioned, in a clearer light. Whether an object is enantiomorphic or not may depend, in part, on its shape; spheres are homomorphs, even in Euclidean space. It also depends on the connectivity of the space, standardly, in global terms. But, what differentiates a thing which is an enantiomorph from one of its incongruent counterparts is a matter of how it is *entered into the space* – how we cut the hole for it in the white strip. Whether we call a knee in an orientable strip a left knee or a right is wholly conventional; it does not really differentiate the knees themselves at all, but simply marks a difference in how they are entered.

The idea of entry is a metaphor, clearly. It springs from our ability to manipulate the knees in three-space, and turn them there so that they fit now one way, now another into the model space. Once in the model space they lose that freedom or mobility and are left only that which determines enantiomorphy. It is not easy to find a way of speaking about this which is not metaphorical. But a very penetrating yet not too painfully explicit way of putting the matter is Kant's own, though I believe it to be still a metaphor. The difference between right and left lies in different actions of the creative cause.

8 Can geometry explain things?

There is a style of relationist argument which we have not yet said anything about, though we promised to do so in Chapter 1 at the end of §2.[1] Furthermore, it may look rather as if there is nothing we can say about it without sawing off a branch we need to sit on. In explaining how handedness is tied up with space in general, as a unity, we said roughly, this (p. 35): a hand is homomorphic if it can be moved into its own reflection and enantiomorphic if it cannot. Now this statement

We pointed out, in Chapter 1, that the two sorts of existential relationism spill over into one another. It is interesting to see Lawrence Sklar speaking of 'the lawlike features of collections of such relations', in Sklar (1974), pp. 281, 285, 287.

is already halfway towards what some relationists see as a perfectly adequate way for them to deal with the problems of left and right. Once we begin to speak of how things could move and of how they would be related to things that could have existed, don't we give the relationists all the licence they need?

This is what Lawrence Sklar thinks (in Sklar (1974)). If we allow relationists the freedom, which they standardly assume, to talk of possible objects and possible continuous rigid motions, then handedness and orientability can be explained perfectly well. The '*real* substanti- valist objection to relationism' is twofold (Sklar (1974), p. 288): either (i) it is nonsense to talk about possible things and possible motions, or (ii) to talk of what is possible is to speak indirectly of 'some actual substantival entity and its actual properties'. These objections have not yet been considered. Remote questions like those of orientability and handedness have no special relevance to the 'real' objections. The crucial battle is over much broader issues, like the relationist's ability to deal with empty space. Hands and global properties of space are just beside the point.

I agree that the relationist sees his talk of what is possible and what not as indispensable for his enterprise. I agree that the standard abso- lutist objections are to this talk and they take a form something like the one Sklar says they take. But they have certainly not been seen as conclusive objections.

Nevertheless, though the main thrust of my paper was directed else- where, we added something to the standard debate. It was this. I might have been born in the same birth as another. To put it in Sklar's style, I have a possible relation to a possible twin. But no property related to twinhood accrues to me for that reason – I am not called a twon, or whatever. Yet all right hands do seem to share a property just if incon- gruent counterparts of them are possible. They are enantiomorphic, handed. The relationist needs not only to assure us that certain spatial relations to certain things are possible. He needs to tell us why we ignore possibilities in general, yet take note of these possibilities in just the way he says we do.

But, to come back to the traditional debate, I do not really think that we can deny relationists their use of modal statements. In particular, the operator 'it is possible that' is useful, to say the least; indeed we can hardly get on without it (see Nerlich (1973), especially pp. 262–8). So one line of argument, which is open to Sklar's substantivalist, is closed to me. But the second line of argument, as Sklar sketches it anyhow, looks quite implausible too. I believe that it is certainly false that 'possibly' generally relates to some 'actual substantival entity and its actual properties'. The two uses of modal expressions discussed in the paragraph after next have nothing at all to do with actual things and their properties.

Nevertheless, this second style of argument against relationism can be developed, much as Sklar says, into a powerful objection, provided that one applies it to a property like orientability. Sklar's way of using modal expressions somewhat obscures this, as I will try to show. He speaks throughout of possibilia: possible objects and possible continuous rigid motions. Now, unlike Sklar, I do not think that a relationist can afford to quantify over motions (i.e. refer to them), possible or not. Any absolutist whose blood is even faintly red, will insist that continuous rigid motions are a kind of spatial entity. They are paths, continua of points, and space itself is nothing but their sum. But, quite apart from running this risk of a warm welcome to the absolutist camp, we ought not to use the adjective 'possible' when we might use the sentence operators 'possibly' or 'it is possible that'. The operators make issues very much clearer. So a relationist should analyse (or reduce) the expression 'orientable' in some such style as this:

O: It is possible that there are two hands a and b such that a is beside b, but it is not possible that a and b move rigidly with respect to c and d etc. so that b will stand to c, d etc. as a now does.

Here, we do not quantify over motions and we avoid 'possible' as an adjective in favour of operators on sentences, open or closed.

It is obvious that the relationist still has the job ahead of him. If these modal expressions mean what they ordinarily mean in philosophical contexts, vague though that is, O is just flatly false. I take the ordinary meaning, among philosophers, to be also expressed by 'It is logically possible' or 'it is conceivable'. In that case, the negative clause of O must be false just because non-orientable spaces are logically possible or conceivable. O is false, even if we construe the modals in the common sense of 'For all I know'. For all I know, our space is *not* orientable and I guess that this is so for all Sklar knows, too. So any relationist who wants to stay in the field owes us some account of what his modal expressions mean. There is no straightforward way of taking them.

Now, we cannot argue like this against the relationist's reduction of empty space – or so I said, by implication, earlier. A careful relationist will probably offer us something along these lines.

E: It is possible that there are two objects a and b and $a\,R\,b$

where R dummies for an expression which puts a or b, or both, in empty places (or whatever is required). Now E, unlike O, is true in both the familiar senses just mentioned, I believe. It is certainly true in the first. As I argued before, we boggle at the idea that space might come to an end, whether or not its boundaries coincide with the limits of a body. Any world we conceive of is a world with a space having no boundaries. So the relationist can claim E in one perfectly straightforward use of its modal expression. This has nothing to do with the actual properties of actual things.

Let us return to O. It is not even enough for the orientability of space that O be true. It might be true for any of a hundred reasons that are consistent with a non-orientable space. It might be true because it is impossible to move objects relative to others, or because it is impossible to move them rigidly. We can look for less perverse reasons than these. It might really be the case that O is true because, although our spacetime yields only a finite non-orientable space for every projection out into a space and a time, still, for all of these projections, the universe has a finite time-span, too short for circumnavigating the space in the limits imposed by the speed of light. So even if we grant O true in some sense, the relationist must make sure that this is the sense he means.

Sklar allows reference to and quantification over possible continuous rigid motions as 'perfectly acceptable relationist talk'. I think this obscures at least this last objection. For, if we speak of possible motions this fixes our gaze on the motions (the paths) as what we must look to. So we forget about the speed of light, in my earlier case, because there are the relevant paths in spacetime. This takes some of the heat off 'possible', over which the charge of gross circularity looms very black indeed in the preceding paragraph. This might even be proper, if we take motions as entities. But it's as plain as the nose on my face, that this commits us to absolutism.

It is not hard to make the charge of circularity stick if the relationist talks as I say he should. One way of doing it is to see modal ideas as Kripke does. We assume a range of worlds. Some are accessible to others in virtue of resemblances among them, expressed in the true necessary sentences of the modal language. For example, granted the standard philosophical sense of modals mentioned before, all the worlds accessible to ours resemble each other in having non-pathological spaces. Different senses of modal expressions correspond to different relations of access, that is, different respects of resemblance among worlds. Which worlds are in the range of a modal operator taken in a given sense depends on respects of resemblance among worlds. What we want the relationist to tell us is which range of worlds his operators cover; that is, how his worlds resemble each other. It is clear what he has to reply. He must say that the worlds at issue are just those alike in having orientable spaces. That is circular analysis and no reduction. I think the relationist account of empty space is no less circular and what I have said makes it obvious enough, no doubt, how I would try to pin that down. But it is not open to the same lines of attack which make the account of orientability so blatantly faulty. There can be little question that the relationist analysis of empty space is much more ingratiating, if not more correct, than his account of handedness. However, rightly or wrongly, in the earlier sections, my fire was directed another way. I should like to make it clear which way that was.

I see the argument just given as directed against the relationist's account of what it means to say that hands are handed and space is orientable. Where the account is faulty and why, seem not very abstruse questions. But the relationist might still save the day if he can show that any actual arrangement of things in space could determine its orientability and limit or empower how things might be moved globally. I threw out this challenge in the last section and offered suggestions as to why (as I think) it cannot be met. Postulating objects never strictly determines even whether they are spatially related or not. Postulating relations among them (specifying how to enter them in a space) never determines global properties of the space. This remains so, even in the context of a theory like General Relativity, as Sklar himself has lucidly observed. (See Sklar (1975), p. 75.)

By contrast, the orientability of space does determine the handedness of hands, for it determines which paths there are in a space which a hand might take. It is a genuinely explanatory idea. This aspect of my argument was not sufficiently clear, perhaps because it is not easy to take orientability seriously as an explanatory concept. It does look rather abstruse, especially when we see it as intended to apply to a weird thing like space as a whole unitary entity. But recourse to the concrete models I used make it clearly a simple, concrete property when ascribed to material things. Among two-spaces, I spoke of table tops, cylindrical and Moebius strips of paper, Klein's (glass?) bottle and so on. It is easy to see how these things differ in the relevant ways. They differ in shape. The Moebius strip is like the paper cylindrical strip, except that it has a twist in it. Orientable spaces, like the two-spaces of the plane and the sphere, clearly differ in shape in many ways.

So, clearly, do non-orientable spaces, as Klein's bottle and elliptic two-space show. In each case we see, from the shape of the space, whether asymmetrical things in it will be handed or not. It is the shape which carries the primary load of explanation. Orientability is a disjunctive property, like being coloured (red or green or...). It is a very general topological aspect of shape, like having a hole. Nevertheless, it is a concrete, explanatory idea. So I claim this advantage: I can explain handedness simply by relating the shape of hands to the shape of the space a hand is in. It strikes me as a very simple and direct explanation, given that we can speak of the shapes of spaces.

The kind of explanation which the shape of space offers us is not really causal. Perhaps that is a main reason for dissatisfaction with it on the part of relationists. But, at this point, the offer of a geometric explanation rather than a causal, mechanical one looks too vulnerable to the charges of being strange and, essentially, empty. The idea of the shape of space is nebulous, remote, abstract. We must give it substance and familiarity. Explanations from the geometry of space are too different from our common paradigms of explaining to draw any strength from work they might do in a variety of other connexions. What else *can* the shape of space explain? Unless we find it other roles besides its part in handedness it will simply perish from isolation and remoteness from the familiar intellectual civilisation. Let us begin the long task of showing just what a shape for space means, in detail, and of showing its power to explain the world to us.

3 Euclidean and other shapes

1 *Space and shape*

Naively, we think that left and right hands differ in shape. But we just saw that there is actually no intrinsic difference of this kind between the things themselves; there can only be a difference in the way they are entered in a certain kind of space. Hands can differ if their containing space is orientable. If it is not they will all be alike however we enter them in it. We found this out by looking at the spaces defined by a paper strip from outside it. A paper cylinder is orientable, but if we cut it and twist it we can change it into the non-orientable Moebius strip. It alters what things do when they move in the space defined by the strip. This looks as if we can say that the difference in the spaces is a difference in their shapes. To ascribe a shape to something surely requires that it must be a particular, individual thing. That was just the suggestion I made about space in Chapter 1: space is not a general, abstract system of relations but an individual, a thing. That was to be its role in ontology and that was what we hoped to make some sense of. But making sense is just the problem. Can we properly speak of the shape of space?

Now it seems all very well to say that space has a shape when we can regard it as defined by a strip of paper, the surface of a ball or of an arbitrarily far extended table top. We see it from outside in a space of higher dimension and it is visibly twisted, curved or flat from that vantage point. How are we to understand someone who talks of the shape of the three-dimensional space we live in?[1] We cannot get outside the space. We have no reason to think it is embedded in a space of higher dimension. We directly see a shape only if it is a limited subspace which our vantage point lies beyond. Yet though the subspace has fewer dimensions than the space which contains it, it need not have any limits in its own dimensions. Any Euclidean plane is an infinite two-space though it will certainly be limited in three-space. The spherical surface, by contrast, is a finite two-space. But it still has no two-space limits. No point of the surface is an end-point. Unlike the plane, it 'comes back on' itself. The plane and the sphere seem to differ in shape therefore not only

[1] I go on to speak of two-spaces, three-spaces etc. in giving the dimensions of a space. I use 'E_2', 'E_3' etc. to speak of Euclidean two- or three-spaces.

as to how they are limited in container three-space within which we view them, but also in themselves – that is, from inside. That looks as if it might make eventual sense of the idea we want of shape from inside a space.

Let's set ourselves some tasks. First, we want to identify a geometric sense in which space has a shape from inside. That needs an excursion into geometries other than Euclid's. I offer the reader three guided tours, so to speak. I begin the survey of Euclidean and other geometries with a sketch of their histories, which suggest a crucial role for the concept of parallels. This section is short and easy. Next, §§3–6 develop the idea of the intrinsic shape of a space. This is rather geometrical and rather detailed, but pictorial and fairly easy. The idea of the shape of space is central to the philosophical argument of the book and the reader ought not to be content with a less complete grasp of it than these sections provide. Thirdly, there are four sections on projective geometry. Again, these are rather geometrical and detailed. The argument is a bit harder, but still pictorial. Perhaps the going is roughest in §9, which touches on analytical ideas. The reader might skip this though it does tie up some loose ends. Again, there is philosophical point to all this. Projective geometry unifies our view of the geometries of constant curvature (including Euclid's). But it also introduces us to the important concepts of transformation and invariance, which underpin the concept of spatial structure, a key concept of this book. If we are to do any effective work with the concept later (from Chapter 7 onwards), we need to understand just what projective geometry is and what role transformations play in it.

2 Non-Euclidean geometry and the problem of parallels

Let us look at the history of non-Euclidean geometries.[1] Euclid began his Elements (1961) by distinguishing three groups of initial statements: there are *definitions* of the various geometrical figures; there are *axioms* supposed to be too obvious to admit of proof; there are *postulates* which admit of no proof from simpler statements, but which are not taken as obvious. Euclid saw these initial statements as providing the foundations

---

[1] This section also comes closest to a quick view of the non-Euclidean geometries as rival axiomatic systems. For a more comprehensive account and for detail on the subject, the classic texts are Bonola (1955) and Somerville (1958).

for a deductive system of geometry. This insight was one of the great path-breaking ideas in Western thought. The path has since been broadened and smoothed into an arterial highway through our intellectual landscape. Some features that Euclid valued strike us, now, as ruts or hillocks; they no longer appear in deductive systems. It is not fruitful to distinguish between axioms and postulates: axioms are now seen as sentences containing more or less familiar general terms. They are chosen severally, because they seem clear and probable, corporately, because they seem mutually consistent and fruitful. Usually, some of the general terms in the axioms will be substantives (for example 'line' or 'set'). These will tell us which domain of study we are making systematic. Others will not be substantives but predicates and relations (for example '...intersects...'). These will tell us which features of things in the domain are under investigation. There will be some definitions in the system but no attempt is made to define the primitive terms that occur in modern axioms. These carry the weight of linking the system to the things it is about. If these primitive terms are left undefined we can reinterpret them as true of other domains and other features of the things in them, without dislocating the proof theory of the system. Our later models of non-Euclidean two-spaces will make use of this fact. Finally, we can classify the axioms of Euclid's geometry and those of other geometries into five main kinds: axioms of connexion, of continuity, of order, of congruence and of parallels. Many different geometries will have axioms of the first four kinds in common.

Euclid made an interesting remark about parallels in his fifth postulate. Putting the matter simply, what he said amounts to this:[1]

'Through any point not on a given line there exists just one line which is both in the same plane as the first line and nowhere intersects it.'

Even the early commentators on Euclid singled out this statement as specially doubtful and bold. But they rightly felt that something like it was needed if geometry was to get far. In short, they kept the postulate because it was fruitful. What caused their uneasiness, presumably, was that the postulate generalises over the whole infinite plane. That clearly goes beyond any experience we have of lines and planes, since our observations are confined to limited parts of these spaces. But the statement grips the imagination powerfully because it is very simple. Rival hypotheses about parallels were invented later, but Euclid's simple insight was so dominant that the rival theories all originated in

---

[1] This form of the postulate is called Playfair's axiom. See Somerville (1958), Chapter 1, §4.

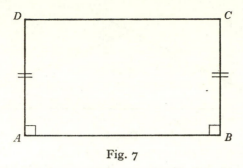

Fig. 7

attempts to reduce every alternative idea to absurdity, so as to prove Euclid's postulate indirectly.

Gerolamo Saccheri's influential work was devoted to this end. He exhausted the rival possibilities by taking a quadrilateral $ABCD$ with two right-angles at A and B and with sides AD and BC equal. He could prove angles C and D equal without the help of Euclid's parallels hypothesis. But he could not prove that each was a right-angle. One way of identifying all the hypotheses in the field, then, is to assert the disjunction that these angles are both acute, both obtuse or both right-angles. So Saccheri and later writers speak of the hypothesis of the acute angle, the hypothesis of the obtuse angle and the hypothesis of the right-angle. (Somerville (1958), Chapter 1, §8; Bonola (1955), §§11–17.)

It was shown early that the hypothesis of the obtuse angle entails that straight lines have only a finite length. This seemed encouragingly absurd,[1] so efforts to demolish the hypothesis of the acute angle went forward with some optimism. But the rewarding breakdown never appeared. Gauss, Lobachevsky and Bolyai each got the idea, independently, that a consistent, intelligible geometry could be developed on the alternative hypothesis. Gauss never published his work though it was found among his letters and papers, so Lobachevsky and Bolyai count as the founders of the subject. Taurinus took the first steps towards a trigonometrical account of the acute angle geometry. He found he could use the mathematics of logarithmic spherical geometry, the radius of the sphere being imaginary. That is, the radius was given by an imaginary (or complex) number; one that has $i = \sqrt{-1}$ as a factor. This introduces a pervasive constant k, related to this radius, into the mensuration. In this geometry, there are two parallels, p_1 and p_2, to a given line q through a point a, outside it. Here neither parallel inter-

---

[1] That lines are infinite was an axiom.

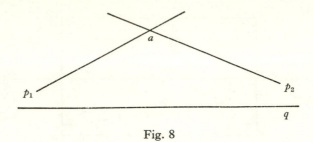

Fig. 8

sects the other line, but approaches it asymptotically[1] in one direction of the plane. The hypothesis of the obtuse angle entails that every line through a intersects q in both directions. Its trigonometry also gives rise to a pervasive constant k. (Somerville (1958), Chapter 1, §11; Bonola (1955), §§36–8.)

This glance at rival axiom sets for geometry highlights the role of axioms of parallels. The new axioms entail that both versions of non-Euclidean geometry have no similar figures of different sizes. The acute angle theory makes the angle sum of a triangle always less than two right-angles, but this 'defect' depends on the area of the triangle and the constant k. Larger triangles have greater defect and the size of other figures also affects their shape. A similar result holds in obtuse angle geometry, but with the angle sum always in excess of two right-angles. However, to characterise the space by its parallels structure is to talk about it only globally. We say something about line intersections in the whole plane, for example. Yet there is reason to think that spaces which are alike in this global way may be more significantly different overall than they are similar. Further the parallels character does little to make sense of the idea of a shape for the space, even though it is global in scope. It turns out that a space may be Euclidean in some regions but not in others; it is then said to be variably curved. That sounds more like an idea of shape. It seems that we have not dug deep enough. Let us get a more general and a more theoretical look at spaces from a new angle: space curvature.

---

[1] Asymptotes are curves which approach one another in a given direction without either meeting or diverging however far produced.

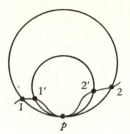

Fig. 9

3 Curves and surfaces[1]

Think, to begin with, of spaces we can stand outside. Then three simple ideas are clearly basic to the theory of curvature: (1) in general, spaces vary from point to point in how they are curved; (2) the simplest curved spaces are circles, being one-dimensional and uniformly curved at each point; (3) small circles curve more sharply than big ones, so we can make the reciprocal of its radius a measure of the curvature of a circle. Now think of any arbitrary one-dimensional continuous curve in the plane. We can measure how much it is curved at each point if we can find a unique circle which fits the curve near the point better than any other circle does. It is called the osculating circle and it is picked out by a simple limiting process. Take a circle which cuts the curve at the point p in question and in two others. Move the other points, 1 and 2, as in the figure, arbitrarily closer and closer along the curve toward the point in question. This generates a series of circles which has a unique circle as its limit. The curve at the point has the same curvature as this circle. We simply take the reciprocal of its radius. One-dimensional curves in three-spaces are a bit more difficult,[2] but we can ignore them. Let us turn directly to curved surfaces.

The problem of defining the curvature at a point for a surface or for a space of still higher dimension, is more complex. We want a strategy which will let us apply the ideas that work so simply in the case of plane curves. We can find it by thinking about various planes and lines which we can always associate with any continuously curved surface. All the curves in the surface which pass through the point have a tangent in space at the point. It is remarkable that all these lie in a plane. So, for

---

[1] A fuller geometrical treatment, but still introductory, may be found in Aleksandrov *et al.* (1964), Chapter 7 and Hilbert & Cohn-Vossen (1952), Chapter 4. It covers the material in §§3–4.

[2] But not much. See Aleksandrov *et al.* (1964), Chapter 7, pp. 74–7.

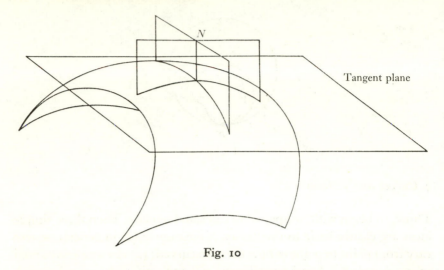

Fig. 10

every point in a curved surface, there is a tangent plane to the surface. Given the tangent plane, we can find a unique line which is both perpendicular to the tangent plane and which passes through the point on the surface. This is the normal N to the surface at the point. Now, consider any plane which contains the normal (as in fig. 10). It will be perpendicular to the tangent plane and will cut the surface in a section which contains the point. The section will give a continuous curve in its own plane and we can settle the curvature of this curve at the point in question exactly as we did before. The curvature of all these sections will describe the curvature of the surface at the point completely.

But things get simpler at a deeper level. First, choose whichever direction of the normal pleases you as its positive sense. Then call the curvature of the section positive if it is concave round the positive direction of the normal and negative if it is concave round the negative direction of the normal. If the plane section is a straight line the curvature of the section is zero, of course. Now these normal section curves will usually have different degrees and signs of curvature. Gauss proved that unless all sections give the same curvature, there will be a unique maximum and a unique minimum section of curvature *and that the sectioning planes containing them will be orthogonal (perpendicular to one another)*. This result greatly simplifies the whole picture. It gives us two principal curvatures, k_1 and k_2. These two curvatures completely determine how the surface is curved at the point, since we can deduce the section in any other plane, given its angle with respect to the principal planes.

4 Intrinsic curvature and intrinsic geometry

But the principal curvatures are still just a means to yet deeper ideas about curvature. We can use them to define a mean or average curvature, though this is not of direct concern to us. It is a useful concept in the study of minimal surfaces, for example. If we throw a soap film across an arbitrary wire contour it will form a minimal surface – one of least area for that perimeter. But we do not want to ponder this average of the principal curvatures, but rather their product. It is called the Gaussian or intrinsic curvature, K, of the surface at the point.

$$K = k_1 k_2.$$

If k_1 and k_2 have the same sign then their product K will be positive and the surface near the point will curve like a ball or a bowl. If k_1 and k_2 differ in sign, the product K will be negative and the surface near the point will curve like a saddle back. If either k_1 or k_2 is zero, K will be zero and the surface is not Gaussianly curved at the point, but can be developed from the plane.

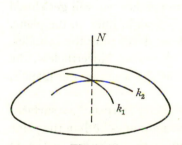

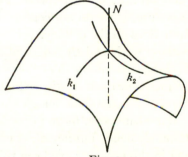

Fig. 11.

$+N$ upward, k_1-, k_2-; $K+$.
$+N$ downward, k_1+, k_2+; $K+$.

Fig. 12.

$+N$ upward, k_1-, k_2+; $K-$.
$+N$ downward, k_1+, k_2-; $K-$.

The idea of developing the plane is intuitively very simple. It is just a matter of bending, without stretching, a plane surface into some other surface. We may also cut and join to get what we want. For example, you can develop a cylinder from the plane by cutting out a rectangle, rolling it into a tube and joining the appropriate edges. The surface is now curved in three-space, in an obvious sense. But what will its principal curvatures be? They will be the same everywhere. Given any point on the surface, one plane through it will intersect the cylinder in a line (i.e. of zero curvature) and other planes will give sections either all positive or all negative, depending on how we picked the direction of the normal. Whatever direction we choose, the linear section will be either the minimum or the maximum of all the curvatures. The section which gives us the circle as its curvature must therefore be the other principal section. So, for every point, the Gaussian curvature will be zero, since everywhere one principal curvature is zero. Gaussian curvature is not the obvious sense in which the cylindrical surface is curved in three-space. Now, not every continuous development of the plane will roll it up into a cylinder. We may get something like an arbitrarily scuffed, undulating or even rolled-up carpet.[1] But however variably curved up it is, in the ordinary sense, we can approximate it as closely as we please with a series of strips from parallel cylinders of varying radius. So the Gaussian curvature at each point on the arbitrarily scuffed carpet is the same as on the cylinder; that is, the same as on the plane. The curvature is everywhere zero.

Let a patch on the cylinder be any contractible space on the surface; it will be bounded by a single closed curve. Now the geometry intrinsic to any patch on the cylinder or the carpet is exactly the geometry of a patch on the Euclidean plane. That is the significance of zero Gaussian curvature. If we do not confine ourselves to patches, we will get closed finite lines on the cylinder corresponding to straight lines on the plane, because we joined the edges of a rectangle in developing the cylinder. Euclidean geometry does not allow that, of course. Nevertheless, the geometry intrinsic to the cylinder counts as the same as the plane's, because it is the same for each patch.

But what does it mean to speak of the geometry intrinsic to a surface? Infinitely many, but not all, of what were straight lines both in the plane and in three-space have been rolled up into circles, or other curves, in order to make the cylinder. They are now curves of three-space. The same is true for the general case of the scuffed carpet. But if we look at all the curves in these developed surfaces, it is obvious that the former straight lines can still be distinguished from the rest. Each is still the shortest curve *contained wholly in the surface* between any two points that lie on it. Having agreed that they are curves of three-space, we can hardly call the 'lines' straight without running a serious risk of ambiguity. So they are called 'geodesics' instead. We can ferret out geodesics in surfaces much more complex than those we can develop from the plane. In general, a curve is a geodesic of a surface if, no matter how

[1] A real rolled-up carpet will also bend in the middle, whereas the plane would not. This is because real carpets also stretch a little.

small the segments we take out of it and however we pick them out, the curve traces the shortest distance in the surface between neighbouring (sufficiently close) points.[1] So it is a curve made up of overlapping shortest curves between nearby points all within the surface. On the sphere, geodesics are great circles or their arcs.[2] For any two points of the surface, the geodesic lies on a great circle joining them. In general, pairs of points cut great circles into unequal arcs. But both count as geodesics since they are made up of overlapping curves of shortest distance between neighbouring points. There are geodesics across arbitrarily and variably Gaussianly curved surfaces, such as mountain ranges, too. Now, to return to the geometry intrinsic to the curved surface, it is just the geometry given by the mensuration of its geodesics.

Perhaps it is getting clearer why the surfaces developed from the plane should keep its intrinsic geometry. Although we bent curves, we did not stretch any. Bending means leaving all arc lengths and angle measures unchanged. So the distances between points in the surface is still the same, *measured along geodesical curves*, though not if measured across lines of three-space outside the surface. In surfaces more complex than the plane, where neither principal curvature is zero at every point, it may still be possible to bend the surface. In these cases, surfaces which can be bent into one another are called applicable surfaces. Obviously, the principal curvatures will change when we bend the surface about. But it is a remarkable fact that their product, the Gaussian curvature, does not change for any point on the surface. This means that surfaces of different Gaussian curvature cannot be transformed into one another by bending, but only by stretching. If we stretch the surface we change arc lengths and angles; this destroys the whole geodesical structure in one area or another. It clearly changes the intrinsic geometry of the surface. So different Gaussian curvatures mean different geometries for the surface as this is given by the geodesics set in it. The product of the principal curvatures reveals something more deep-seated in the space at the point than the principal curvatures alone would tell us. It tells us what the surface is like from inside.

[1] A more careful account of geodesics is given in Chapter 9, §6.
[2] A great circle contains the point diametrically opposite any point on it. Circles of longitude are great circles, but only the equator is a great circle among circles of latitude.

5 Bending, stretching and intrinsic shape

Each principal curvature describes, primarily, how the surface is bent with respect to the higher dimensional space outside it. Clearly, that concerns the shape of the two-space but from outside as it stands in three-space. It is a question of the limits of the surface in the containing space. But, as we saw, the surface need not be limited in its own dimensions, whether it is infinite like the plane or finite like the sphere. So there is clearly a more general, and clearly a deeper, idea of the shape of the surface which is captured by the Gaussian curvature. A carpet is rightly described as a flat surface, even if it happens to be scuffed or rolled up either right now, or rather often, or even always. A hemisphere can be quite freely bent with respect to its surrounding space, yet there is a general notion of its having a shape which does not change when we bend it. Consider the familiar and, I think, quite uncontrived example of the human shape. The body surface is a finite, continuous, unlimited two-space (if we overlook the holes, for the sake of simplicity and, perhaps, decency) which we regard as a quite particular shape. It has a very intricate, variable, Gaussian curvature. But think of the wide array of postures and attitudes in which we quite ordinarily identify this single shape! Thus we find the same human shape in a person bending, crouching, running, sitting, standing to attention or with arms akimbo. But men differ in shape from women, fat people aren't the same shape as thin ones, and so on. (The example is not quite perfect, of course – the skin stretches a bit, which allows for more mobility, but the inner bones are rigid, which allows for less. I cannot see that these blemishes affect the point, however.) We can equally well distinguish a general shape and a particular posture for it on behalf of two-spaces generally. Shape is intrinsic to the surface, posture is not. Gaussian curvature tells us everything about shape, but in most cases, nothing much about posture but the range open to it. Gaussian curvature is inside information.

Given the main aim of this chapter – to make sense of the idea that space has a shape – that is an important conclusion.

Clearly enough, if we can tell the intrinsic geometry from the Gaussian curvature, then, conversely, we can get this curvature out of the geometry as measured from inside the space. We need to look at the geometry in patches round each point and discover how the space is intrinsically curved everywhere. That will tell us its shape without our needing to discover its posture first by hunting out principal curvatures. Suppose we start with the geometry of some three-space and look at the measurement of distances between points along geodesics in various volumes (hyper-patches) containing different points. We may find that the geometry varies from point to point, and we can assign a Gaussian curvature to the space at each point as the measurements suggest. It is easy to see in this case that the curvature neither tells us whether there is a containing four-space in which the three-space has a posture, nor, of course, just how it is limited and bent (postured) if there is a four-space. The sense in which a three-space has an intrinsic shape is precisely the same as the sense in which a two-space like the surface of the human body has one, independently of how it lies in containing three-space.[1] So a space is an individual thing with the character appropriate to individuals; it has a shape. We still need to give this concept some percepts, for three-spaces, at least in imagination. But telling how a sharply curved three-space would impinge on our nerve endings if we were inside it must wait till the geometric stage is set more fully. We can quickly look at some suggestive and interesting two-spaces – from outside, however.

6 Some curved two-spaces

There are surfaces in E_3 which provide very simple examples of what we have been discussing. I have already mentioned geodesics of the sphere and it is fairly obvious that the spherical surface is a two-space whose Gaussian curvature is everywhere constant and positive. Though the saddle back of fig. 5 is a negatively curved surface it is not constantly curved. If the curvature of a saddle back is to be everywhere negative it must vary in degree. However, the pseudo-sphere is an example of a two-space in E_3 which has constant negative curvature (see fig. 13), though it has a family of finite geodesics on it and a singularity at the central ridge. But any patch on the pseudo-sphere outside the singularity has the geometry of constant negative curvature. Unluckily it is not so enlightening about negative curvature as the sphere is about positive curvature. An interesting example of a simple space of variable curvature is given by the torus, the surface of a doughnut or anchor ring (see fig. 14). The curvature is clearly positive at point *a* and negative at point *b*. Therefore it must be zero somewhere, namely at all the points

---

[1] The geometry of three-space curvature defined from inside the space is more difficult than this. It will be explained in Chapter 9.

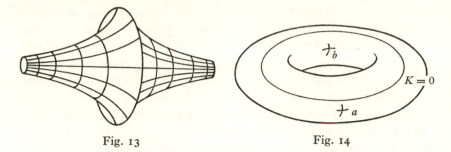

Fig. 13   Fig. 14

of either of the two circles where the torus would touch any flat surface on which it rested. One of these circles is shown in the figure.

How do these ideas connect with the earlier investigation of parallels? You will not be surprised to hear that the Gaussian curvature K can be related quite directly to the trigonometric constant k discovered by Taurinus and pervasive in both acute angle and obtuse angle geometry.

On the sphere, every pair of great circles intersects (twice, in fact, in polar opposite points) so there are no parallels on the sphere, just as there are none in obtuse angle geometry. Every triangle on the sphere (composed of arcs of great circles) exceeds two right-angles in its angle sum, the defect being more the smaller the ratio of the triangle's area is to the total area of the sphere (that is, roughly, the larger the triangle). On the pseudo-sphere the 'horizontal' geodesics of the figure obviously approach each other asymptotically in the direction away from the ridge, but this singularity prevents a simple representation of the pair of parallels permitted in the geometry of Bolyai–Lobachevsky. On the torus, there is a sort of Euclidean parallelism. We can find a geodesic on the torus such that, given a point outside it, there is just one geodesic that nowhere intersects the first – in fact, we can find an infinity of such geodesics. But how different this surface is in its intrinsic geometry from what we find on the plane!

Within this rather extensive view of spaces of continuous curvature, it is very useful to single out those which projective geometry deals with. Let us turn to that.

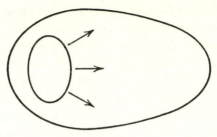

Fig. 15

7 Perspective and projective geometry[1]

Spaces of constant curvature lie at the centre of geometry's field of interest. Two reasons explain this. First, there are needs of physical realism. Helmholtz (1960) and Lie (Somerville (1958)) established that only spaces of constant curvature allow the free mobility of all objects. Clearly, a thing in the two-space defined by an eggshell cannot move freely about it without changing its shape. If the curve in fig. 15 were a paper patch lying on the eggshell, then it would tear if we tried to fit it onto flatter areas near the middle of the shell or it would wrinkle if we tried to fit it onto the more acutely curved regions near the 'pointed' end. If it were to slide along the space then it would either have to deform like an elastic thing (stretch, not merely bend) or it would resist the motion. Our experience of how things move in space suggests that they are very freely mobile. Until General Relativity, at least, constant curvature seemed necessary for any space if it were to be regarded as perhaps physically real. A second, perhaps more important reason, for interest in spaces of constant curvature is the unified view of them which projective geometry presents to us. The subject is a thing of beauty in itself and will give us some concepts we can put to work later. Let us take a quick look at it.

If an artist draws a perspective picture of a scene, what appears on his paper are recognisable likenesses of things. Obviously enough, that is because the figures on the paper share some geometric characteristics of their subjects. But the illusion of perspective also depends on some departures from the shape of the objects portrayed. What must the artist alter and what must he preserve? Think just of the simple problem

[1] For a detailed, introductory treatment, see Aleksandrov *et al.* (1964), Part 2, Chapter 3, §§11–14; Adler (1958), Parts 2 and 3; Bonola (1955), appendix 4; and Courant & Robbins (1941), Chapter 4, give synopses of the material. These cover the material in §§7–9.

of putting on a roughly vertical canvas a perspective representation of shapes and patterns on a single horizontal plane. The problems of translating three-space objects into flat outlines hold no interest for us at present. The painter is to project one plane onto another, the lines of projection all passing through a common point, his eye.

What happens is that lines on the horizontal plane are projected into lines on the canvas (the projection plane); non-linear curves are projected into non-linear curves (with exceptions we can afford to ignore). Any point of intersection of lines (or curves) is projected into a point of intersection of the projections of the lines (or curves); if two points are on a line (or curve) then the projections of the points lie on the projection of the lines (or curve). But distance between points is not preserved. This is clear from the striking and important fact that the painter draws a horizon in his picture. This is not a portrayal of the curvature of the earth and the finite distance from us of the places where the surface recedes from sight. Even for the horizontal Euclidean plane the artist who puts a perspective drawing of a horizontal plane onto a vertical one must draw a unique horizon across it. He represents an infinite distance by a finite one.

The role of the horizon brings us back to the question of parallels. Lines which are parallel in the horizontal plane and do not intersect in it will be projected into lines which converge on points that lie on this line of the perspective horizon. Sets of parallel lines that differ in direction will be projected into lines that converge on different points on the line of the horizon. A real line on the artist's canvas represents an improper idea;[1] there are infinitely distant 'points on the plane' at which parallels 'meet' and all these 'points' lie on a single straight 'line' (or 'curve') at infinity. Projective geometry therefore completes the planes it studies with points and curves at infinity. We could never come across such points in traversing the plane and nowhere on it do parallel lines begin to converge, let alone meet. What justifies the addition of improper elements is, first, that sense can be made of it in analysis and, second, that it suggests itself so naturally in the elementary ideas of perspective which we just put through their paces.

But our view is still not general enough. What an artist does is to use

[1] Not to say indecent. I am avoiding use of 'imaginary' which is the obvious word to choose, as it bears a quite different sense in the analysis of projective geometry, where one has to deal with the algebra of imaginary and complex numbers. The line at infinity is represented by real values of homogeneous coordinates. It is real in the analytic sense. But I wish to leave these techniques aside.

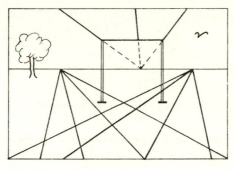

Fig. 16

the space on his canvas above the line of the horizon to portray things that do not exist in the horizontal plane. He pays the price of representing only that part of the horizontal plane which is in front of him (fig. 16). Since he looks in only one direction the projective apparatus is really just by rays (half lines) from the eye 'through' the canvas to points on the plane. But if we both take the whole pencil[1] of lines on the point (his eye) and give a geometer two infinite planes to play with, we will get an unrestricted view of what plane projective geometry is about. What will be mapped onto points above the line of horizon h will now be points in the horizontal plane behind the plane of projection, as fig. 17 illustrates. It also shows us that, not only will the improper curve at infinity map into a real line of the projection plane h but, conversely, the real line L of the horizontal plane will map only into the improper line at infinity of the projection plane. For the plane through P and L runs parallel to the projection plane. So the line at infinity is generally changed in projective geometry and so, therefore, are parallels.

On the complete planes of projective geometry (or on *the* projective plane, as we will soon be entitled to call it) we can say that every two lines join on just one point and that every two points join on just one line. For now even parallel lines join on points at infinity. This gives rise to a very elegant and important principle of duality between points and lines for the projective geometry of the plane. Each theorem about points has a dual theorem about lines and for every definition of a property of points there is a dual definition of a property of lines. So it is useful to speak of points 'joining on' lines and of both pencils of lines and pencils

---

[1] The pencil of lines on a point is just the infinite class of lines that contain it. We can also speak, dually, of the pencil of points on a line as the class of all points contained by the line.

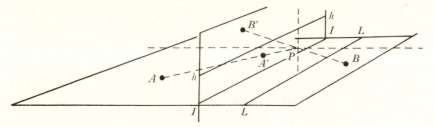

Fig. 17. Points A' and B' on the projection plane are the images under projection through P of the points A and B on the horizontal plane.

of points. In the projective geometry of three-spaces, planes and points
are dual (and lines are self-dual, trivially).

8 *Transformations and invariants*

We want to continue to steer wide of the holy ground of analytical
methods of investigating space. Still, we can get a more general and
much deeper insight into the projective geometry of n-spaces[1] if we
ponder the elements of the theory of transformations of a space. There
is no doubt that this theory draws its strength from the ease and accuracy
with which it can be pursued analytically. But we can still profit from a
qualitative view of transformations. The projective geometry of spaces
of three or more dimensions is the study of a certain group of transformations of the space, and it does not require an outside space across
which projection is carried on. A transformation is simply a function or
mapping defined over the points of any space which carries each point
into some point of the space (usually a different point, of course). In
perspective drawing we map one space (plane) onto another by visual
linear projection. But if we now rotate the projection plane round the
line in which the two planes intersect (I in fig. 17) till both coincide, we
can see that the mapping might now be regarded as carrying the first
plane *into itself*, by mapping points on the plane into other points of that
same plane. We can regard the projective geometry of n-spaces generally
as the study of certain transformations of the space into itself. This is
why one speaks of *the* projective plane. Clearly enough, these transformations are 1–1 and continuous, except for mappings of finite into
infinite points and vice versa. But, as the idea of perspective projection
suggests, the transformations are linear as well. So projective geometry

[1] That is, spaces of n dimensions.

is the study of all those transformations of a space into itself which are linear and (for the most part) 1–1 and continuous; the projective geometry of a three-space is the study of such transformations of three-space into itself.

When an area of geometry is defined by means of the transformations it studies, these are not picked at random: they always form a group. This means that the product of two transformations (the result of following one with the other) must also be a member of the collection of transformations. So projective geometry will describe perspectives of perspectives. Further, for each transformation of the collection there is an inverse transformation which reverses it and is also in the group. It follows that there must be an identity transformation, which changes nothing. It is given by the product of a transformation with its inverse. It leaves everything the same.

Some properties of objects in the space will be preserved whichever transformation of the group we use on the space. But other properties, length and angles in projective transformations for example, will be altered by some members of the group. The first sort of property is called an invariant of the group of transformations; the second a variant property. Projective geometry can be seen as the study of invariants of the projective group of transformations. In fact, every geometry can be regarded as the study of the invariants of transformations of some definite group or other, as we will see in more detail later.

What are the invariants of projective geometry? They include the ones we saw before: lines are invariant, being always transformed into lines. Intersections are invariant: if two points are on one line then the transformed images of the points lie on the images of the line; if two lines are on (intersect at) a point then the images of those lines are on the image of the point. Projective transformations are called collineations for this reason. There are similar invariant incidence properties for points, lines, planes etc. for spaces of more than two dimensions. Distance proper is a variant of the projective group, as our earlier glance at perspective drawing showed. However, a related notion of the cross-ratio of collinear points is invariant. If we take four points $ABCD$ on a line then the ratio $(AC/CB)/(AD/DB)$ (written (AB, CD)) remains the same for all projective transformations of the points.

But as we saw in perspective, the line at infinity is not invariant under these transformations. Nor, therefore, are parallels. The improper line at infinity may easily get mapped into a real finite line on which parallels will be mapped as intersecting. Analytically it is an advantage to let there be improper points at infinity which can vary under projective transformation. In fact, as Poncelet (Adler (1958), §§6.7, 6.8) was first to see, the whole subject takes on a unity and profundity if we also include imaginary and complex points and study the whole complex *n*-space. It then turns out that if we regard the line at infinity (or the plane at infinity for three-spaces) as some kind of conic, and treat it as a special element of complex space, the non-Euclidean geometries of constant curvature, and Euclidean geometry too, can be shown to be different branches of one overarching subject, projective geometry. We can then understand the point of a well-known remark of Arthur Cayley's: 'Projective geometry is *all* geometry.'

9 Subgeometries of perspective geometry[1]

A conic in Euclidean metrical geometry is some section of a cone. Different sections give rise to a variety of curves among which we should notice the ellipse, the parabola, the hyperbola and the straight line, considered as a degenerate conic produced when the sectioning plane is tangent to the cone. But these are metrical ideas and variants in projective geometry; we want a characterisation framed in the right properties. Consider two points, p_1 and p_2, and the pencils of lines through them. (See fig. 18.) Pick any three lines (a_1, b_1, c_1; a_2, b_2, c_2) from each pencil and pair them off at random. Label their points of intersection (a_1, a_2), (b_1, b_2), (c_1, c_2). Then the five points, p_1, p_2 and the three just mentioned, determine a *projective correspondence* between all the lines of the two pencils. The points of intersection of corresponding lines in the pencil generate a curve. This is called a conic since it is fixed by five points and since any line intersects it in at most two points. It resembles the familiar conic of metrical geometry in both these ways. (It is also possible, dually, to generate the same curves by the pencils of points on two lines and a projective correspondence between three pairs of such points.)

It is obvious that if we select some subgroup from a group of transformations, then this smaller set of transformations will change fewer properties of elements in the space. That is, the subgroup will have more invariants. Now, in fact, Euclidean geometry does consist of trans-

---

[1] This section edges a little into analysis. The reader could skip it, but if he really is a student of the subject he needs some grip on the ideas. It unifies Euclidean and non-Euclidean spaces as all part of projective space.

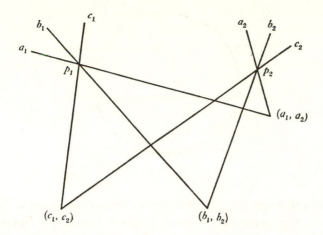

Fig. 18. A correspondence between three pairs of elements from the pencils of lines p_1 and p_2 determines a projective relationship between the pencils. This defines a conic (Adler (1958), §§9.2, 9.9).

formations which are a subgroup of projective transformations. Euclidean geometry is defined by the set of invariants under the group of transformations which are rigid motions and reflections. Since figures of different sizes are similar in Euclidean geometry, the group should include dilations too. How can we select this Euclidean group out of the wider projective group? Select any degenerate conic, any line, to be called the *absolute*. It is taken to be the line at infinity, just as a finite line may be taken as a horizon. Then we select the transformations from the projective group which leave this line invariant (which map it into itself). Intuitively, these will include perspectives from one plane onto a parallel plane, for example. These transformations are none other than the rigid motions of Euclidean space.[1]

But now suppose we select any non-degenerate real conic of the projective plane as the absolute (curve at infinity) and consider the subgroup of transformations which map this conic into itself. Lines of the plane (other than tangents to the conic etc.) intersect it in two points O and U. Given two points on such a line, A and B, we can define the distance between them as the logarithm of the cross-ratio (OU, AB) multiplied by a constant.[2] Thus $d(AB) = c\log(OU, AB)$. When distance

---

[1] Strictly, they are rigid motions in the analytic sense for homogeneous co-ordinates. Thus they include similarity transformations.

[2] Thus the absolute will be infinitely distant from any point inside the conic, according to this metric.

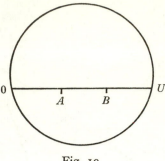

Fig. 19

is defined in this way we get precisely the metric geometry of the rigid motions on the plane of constant negative curvature. Suppose we consider, now, the complete complex plane and choose as the absolute, infinitely distant curve, a purely imaginary non-degenerate conic. This will mean that each real point will be at a finite distance from every other. Since every two real lines join on a real proper point, there will be no parallels. The subgroup of projective transformations which leaves this conic invariant gives us the metric geometry of rigid motions in a plane of constant positive curvature. This is not much help to visualisation, but the non-mathematical reader can no doubt see to some extent how all these metrical geometries come together as subgeometries of projective geometry. We have seen, again, how these spaces differ without slipping into a space of higher dimension to view them from outside.

But what does this have to do with curvature and parallels? The general equation of the conic in homogeneous coordinates[1] is as follows:

$$k(x_1{}^2 + x_2{}^2) + x_3{}^2 = 0.$$

If the constant k is zero, the absolute conic becomes degenerate (which yields Euclidean geometry of zero Gaussian curvature); if $k < 0$, the absolute conic becomes real and non-degenerate (which gives the non-Euclidean geometry of negative curvature); if $k > 0$ the absolute conic becomes imaginary (giving the non-Euclidean geometry of positive curvature). It will cause no surprise that the constant k can be identified with the Gaussian curvature K. We can define parallels as lines which

---

[1] Homogeneous coordinates give three variables for the coordinated plane. There are various reasons why this is convenient. A clear account of them, which makes their use intuitively acceptable, can be found in Adler (1958), §§ 10.2–10.3.

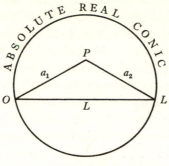

Fig. 20

intersect at infinity, that is, on the absolute conic. The intersecting points are distinct and real if the conic is non-degenerate and real. So there are two parallels to a line from any point outside it, just as required for the geometry of the acute angle in spaces of constant negative curvature (see fig. 20). The points are coincident and real if the conic is degenerate and real (reflection on the perspective projection shown in fig. 17, p. 66 shows that points at infinity both in front of the projection plane and behind it are mapped onto the same line of horizon). This permits one parallel through any exterior point. If the conic is wholly imaginary the points of intersection are also imaginary, so no real parallels are possible. Felix Klein pointed out that this was strikingly analogous to the existence of asymptotes to the hyperbola (two), the parabola (one) and the ellipse (none). So he named the three classes of metrical subgeometries of projective geometry hyperbolic, parabolic and elliptical geometries. Hyperbolic and elliptic geometries are classes of geometries since the curvature of the space may differ not only in sign but also in degree. These conclusions can be generalised to spaces of more than two dimensions, obviously enough. So hyperbolic, parabolic and elliptic geometries for n-space are all subgeometries of projective geometry.

10 More two-space models in E_3

Now we can look for new models of non-Euclidean two-spaces in Euclidean space. Let us take up the one we caught a glimpse of a moment ago (fig. 20). Points and lines within the circle represent real points and lines of projective space and the circle can be taken as a non-degenerate conic giving the points at infinity. It is the absolute. If we measure lengths on the real lines by the logarithm of the cross-ratio between points on the line and the infinitely distant points O and U, we get a model of hyperbolic geometry. In the figure, L is a complete real line infinite in a length according to the distance function we accepted. a_1 and a_2 are distinct infinite rays through the point p which are parallel to L, meeting it at infinity. They approach L asymptotically in terms of the distance function postulated. We can get a different model, having a different interest, by interpreting circles which intersect the absolute at right angles to be lines of the space. We can adopt the same distance function as before. But the Euclidean angles of the figures drawn in this new model will equal the hyperbolic angles which they model. The model is conformal. It portrays the shapes of figures in a useful way.

I said earlier that the axioms about incidence in projective geometry are closely related to the Euclidean ones, except for an extension allowing parallels to intersect on the line at infinity. So the incidence axioms of projective geometry stipulate that two lines join on just one point and this axiom will carry over into elliptic geometry. This is a geometry of constant positive curvature. The earlier model of a two-space within E_3 having constant positive curvature was the spherical surface. But it is clear enough that the surface of a sphere is not an elliptic plane. On the sphere, any two lines join on two points. Further, the dual elliptic axiom of incidence, that every two points join on one line, is false on the sphere: any pair of polar points join on infinitely many lines of the sphere. The trouble is that the sphere is a kind of double space; there is twice as much of it as we need. We can get half of it in two ways: one is simply to identify each point on the sphere with its polar point. An equivalent way is easier to visualise – take a hemisphere (see fig. 21) and identify diametrically opposite points on its equatorial section as suggested by the figure. This provides a good model of the elliptic plane. Lines join on just one point; points join on just one line. Incidentally, this plane is non-orientable, just as the Moebius strip is.

Before I turn to other aspects of the problem, let me rehearse what we have knitted our brows so tightly over up to this point. First, we looked at geometry graphically and historically by thinking about the question of parallels; then we went into the matter of curvature, discovering what it means to speak of the shape of a space as an inside matter given by its Gaussian curvature at all points, and distinct from any posture it may (or may not) have; then we looked into projective geometry to find a deeper unity underlying the sharp differences among the spaces of

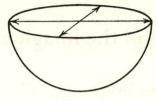

Fig. 21

constant curvature. We will need all this to understand how to grasp the idea of a shape for space in the intuitive sensory imagination. Much of the survey will be useful to us in other ways as well.

The models of non-Euclidean space fall well short of what we want from an intuitive glimpse into these spaces. The geodesics of the two-space of the sphere, elliptic hemisphere and torus are visibly curved in E_3 and the logarithmic length measure used in the model of hyperbolic two-space makes no concessions to vision, despite remarks of Reichenbach ((1958), p. 58). We have made progress on the way, but we are still far from home.

4 Shapes and the imagination

1 Kant's idea: things look Euclidean

The aim of this chapter is to describe what it would be like to live in a markedly non-Euclidean world. We want a description that appeals to the visual and tactile imagination. How would things strike the eye or cleave to the touch in a strange space?

On the face of it, this is an entertaining exercise which we can start with real hope of success. Some effort of abstract thought and some agility of the visual imagination will no doubt be needed. But the task certainly looks possible. The non-Euclidean geometries are consistent. There are detailed analytical descriptions of non-Euclidean spaces, which are consistent if arithmetic is.[1] Looking in another direction, we find that
e can lay down axioms of non-Euclidean geometry and reinterpret their primitive terms as true of certain objects in Euclidean space,[2] as we did at the end of the last chapter. Neither of these gives us the description of visual experience that we want. But they should encourage us to think that we can find it.

Nevertheless, the view that our imaginations are bound by Euclidean limits has some supporters. Kant held that Euclidean three-dimensional space (E_3 for short) had to be the space in which we localise, particularise and differentiate individual appearances (Kant (1961), 'Transcendental Aesthetic'). But what Kant meant by this was something subtle. He meant that the mind was bound to impose E_3 as a form of intuition, constructed by itself to objectify appearances. So Kant did not hold that real, objective, independent space had to be E_3. His arguments would certainly have led him to reject the idea that we could know the structure of any real space *a priori*. He believed only that what we might call merely visual or phenomenal space was necessarily E_3.[3] We have no choice but to concede that real space might well be non-Euclidean. That is perfectly consistent with holding that visual space is just as Kant said. But was he right?

---

[1] For example in analytic treatments of projective spaces: see Adler (1958), Chapter 12.

[2] See also Somerville (1958), Chapter 5; Hilbert & Cohn-Vossen (1952), §§34–9.

[3] But Kant allows no distinction between 'subjective' and 'real' space.

First, what is visual space? On the face of it, there is only physical space in which we see things, the sky and so on. True, we speak of a visual field. But this need not mean more than simply the area or volume which someone can take in at a glance. Still, the idea that there is a unique space of vision, private to each perceiver, and peopled by merely visual objects, has been a common conviction in the history of epistemology. At present, few philosophers would regard it with much enthusiasm, however. What gives the idea some plausibility is that things do not always look as they are. It was supposed to follow that whenever we are aware *that* a thing looks (but is not) so and so, then we are aware *of* something that really is so and so. For example, if a straight stick, partly immersed, looks bent then we are aware of something that is bent. This something can only inhabit a purely visual space and be a purely visual object, since there will usually be nothing in real, objective space that is so-and-so (bent). Kant certainly distinguished appearances from things themselves. We should take Kant's assertion to be about these visual objects and their space.

We have reason to regard these ideas with deep suspicion. Even some of Kant's modern defenders share this distrust. Luckily, we need not now descend into the intellectual quagmire in which these issues have sunk in order to press on without begging questions. We can speak just about how things look on the one hand, and how they are, on the other. The disputations about whether there are (or can be) visual objects or visual space can be sidestepped. Difficulties about what would be involved in admitting such things to our ontology can be dodged.[1] Let us interpret Kant's thesis to be that however things may be in their geometric properties and however space may be curved, the world will always look Euclidean. So let us turn to the question itself, again. Was he right?

2 Two Kantian arguments: the visual challenge

Some impressive men agree that he was. Frege ((1950), pp. 20, 101 f.) is robustly Kantian on the question. P. F. Strawson draws a similar conclusion, if a little less roundly. Carnap ((1922), p. 22) went at least some of the way to the same conclusion. There seem to be two kinds of argument for these neo-Kantian theories. The main argument seems to be just a challenge to visualise a certain kind of example. The second is a

---

[1] For now, but they are discussed later in Chapter 6.

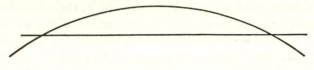

Fig. 22

general explanation of why our limited powers of vision would rule out any gross violation of Euclidean appearances.

The kind of challenge offered by arguers of the first sort is clear enough from these quotations:

'If we restricted ourselves to what can be "imagined" or seen at a glance, then perhaps we should be bound to regard space as Euclidean...It is not clear how we could see at a glance that two straight lines intersect twice: it seems that if both intersections are seen at once, then at least one of the lines must look curved' (Bennett (1967), p. 31).[1]

'Consider the proposition that not more than one straight line can be drawn between two points. The natural way to satisfy ourselves of the truth of this axiom of phenomenal geometry is to consider an actual or imagined figure. When we do this it becomes evident that we cannot, either in the imagination or on paper, give ourselves a picture such that we are prepared to say of it both that it shows two distinct straight lines, and that it shows these lines as drawn between the same two points' (Strawson (1966), p. 283).

The idea is that any attempt to see or visualise the appearance in question would make us bend a line somewhere, or 'cheat' in another way, so the attempt would fail. Certainly, the upper line in fig. 22 is bent. It is far from obvious how we could even begin to set about meeting Strawson's challenge. This particular case seems to have attracted quite a lot of attention.

But it is a queer case. Clearly, we can't do what we are asked to do. It is less clear why we should try. The task comes straight out of the blue with no explanation of what calls for it in non-Euclidean geometry, if anything does. Before we try to deal with it, let us see how and where the challenge arises. Now we saw before that in hyperbolic and elliptic geometries the incidence axioms for lines are the same as in Euclidean space. They are subgeometries of projective geometry. Two lines intersect on just one point; two points lie on just one line. So failure to meet the challenge is perfectly consistent with visualising any non-

[1] But Bennett does not conclude simply that visual space must be E_3.

Euclidean space of these sorts. It is simply a flat geometrical mistake to suppose that if we can't visualise this case, our imaginations are Euclideanly bound. In spherical space, things are different. For each point there is another point linked to it by infinitely many lines. However, this crops up only for a point and its polar opposite point. But then, every line through the one point passes through the other. Only in spaces of rather acute, variable curvature do we get more or less nearby points joined by a few, not too long, geodesics. So, for the visual challenge, there are only two sorts of case that possibly matter. Let us take a closer look at each of them.

3 Non-Euclidean perspective: the geometry of vision

A bit of reflection on the facts of perspective shows that only in fairly well-behaved, fairly regular spaces, could we begin to expect things to look much like they actually are. Suppose we are in a three-space of acute, but variable, continuous curvature, and assume here, as always, that light paths are geodesical. Then the different rays which project seen objects onto the retina will each be variably curved from point to point in its path. But, more importantly, rays from some parts of the object will be differently curved from those from other parts. We will get a continuously distorted retinal image, which will twist further in all sorts of ways as either we move or the object moves. This is quite beside the problem of free mobility for ourselves or the objects, which might cause actual warping in us and our retinas, or in the moving objects we see. The only useful generalisation about vision in such spaces is that things would look very changeable in aspect. It would be very like seeing things through a strong heat haze.[1]

It is more challenging and more instructive to exercise the imagination in fairly simple spaces. Let us take a fairly obvious geometrical approach to the problem. Imagine a small glass screen, like a monocle, in front of the eye. This screen is a small region (a disc) in a plane of the space, and thus a non-Euclidean plane when the space in question is non-Euclidean. We may call this disc the retina. In ordinary experience, much of how things look is determined by what the geometry of their projections onto this retina would be. A tilted coin is projected into an ellipse, for example. We assume, as always, that light paths are linear.

---

[1] Later, I describe experience in a simple case of variable curvature, the three-space analogue of the torus. But we do not get distinct geodesics through pairs of nearby points in spaces like this.

Then it seems reasonable to say, in all the simple projective spaces, that how things look is just the geometry of their projection onto the small disc. We include spherical space in this since its curvature is constant.

Now it is a projective theorem that lines and linear figures in the space will go into lines and linear figures on the retina. Consider just the simplest case where you see the figure 'head on'. That is, where the retina and the face of the seen figure are joined by a line orthogonal to both which passes through their centres. Any large triangle in the space will have a large defect – a departure of its angle sum from 180°. But its projection onto a small disc in a plane of the space will yield a small triangle. It can only have a small defect. It will be very nearly a Euclidean figure. The retinal projections of things *will not be similar to the things. Figures and things will not look as they are, in general. They will look pretty much Euclidean.* This is quite a general argument. It clearly touches all linear, polygonal shapes since any of these can always be represented as a complex of triangles. We need only draw all the diagonals from one vertex. It is worth noticing that this is not a perceptual argument for the conclusion, but a purely geometrical one. It is remarkable that the most natural assumptions about vision – the small plane retina in a space of constant curvature – yields a Euclidean look to everything. Of course, we are not bound to limit our assumptions in these ways.

But let us probe the core of the example given by fig. 22. It arises as a clear problem for vision only in spherical three-space. For only there do we get clear principles that there are distinct lines, each through the same pair of points, and that there is a simple relation between the geometry of a retinal projection and the geometry of the thing projected.[1] In any ordinary circumstances, we will see one of the polar points from some distance away, along a geodesic joining the point and the eye. Then if one of two polar points is in front of the retinal plane and in the visual field, the other must be unseen. The point (or the area) which we are looking directly at will mask it.[2] The only way to see the second point from our viewing position is to turn right round and look in the opposite direction. To get both points projected onto the retinal plane in one look

---

[1] Helmholtz's model for visualising non-Euclidean space is faulty in this respect. He suggests we look at the world in a convex mirror. But one's retina (or retinal plane) still remains Euclidean, so we do not get just what we want. For example geodesics get projected into curves of the retina. This criticism applies to essentially the same suggestion by Eddington (Helmholtz (1960), pp. 661–4; Eddington (1959), p. 11).

[2] If the areas are equal and the shapes the same, the one will exactly mask the other.

we must choose the plane so that both points lie actually on it. This would be a trivial case of projection since the points and their 'images' are the same. But even then, not both points would be seen unless the visually sensitive disc of the retinal plane extends from one point right across to the other. Therefore, the sensitive retinal plane must occupy half of the total (finite) region of the complete plane in spherical three-space. That would be a queer case of vision indeed. We need not wonder if imagination boggles at the task of conceiving it.[1]

4 Reid's non-Euclidean geometry of visibles

Suppose we enlarge the powers allowed to vision by considering one very natural extension. Imagine that the eye integrates whatever it can see by rotating the head as it follows a curve. Make a corresponding extension of what visual geometry is to be the geometry of: instead of asking for the geometry of objects as we see them, ask instead for the geometry of the points, lines and linear polygons made up by this integrating process. Thomas Reid, in 1764, discovered a non-Euclidean geometry in this way, roughly thirty years[2] before Gauss began private conjectures as to alternatives to Euclid. Reid came upon spherical or doubly elliptic (bipolar) geometry in two dimensions; the two-space modelled on the Euclidean sphere.

To understand this we need to look again at the geometry of perspective. Suppose real space is E_3 and visual space retains and integrates what we see when we turn our heads so as to follow a curve or line. Then the geometrical problem is neatly given by features of the pencil of rays (half lines) through a point which represents the eye. To each ray through the eye there corresponds a perceived point (roughly). The rays that project a perceived line will form (a portion of) a plane in which both the line and the eye fall. So let us attend to the whole pencil of rays and planes through the point representing the eye. This gives us a class equivalent to the class of points and lines that may be perceived.

Nothing in this suggests that we should revise our idea that seen

---

[1] Eyes on each side of your head would help. It helps birds with other problems. The reasons why we falter at this hurdle for the visual imagination do not hold much philosophical interest.

[2] This is a debatable period of priority. I judge from a footnote on p. 65 of Bonola (1955). The actual priority seems beyond dispute, though. See Daniels (1972); Reid (1847), Chapter 4, §9.

figures will look Euclidean.[1] It suggests a new way of approaching the whole problem of visual geometry. The linear edges of quite different things widely separated in E_3 may be projected by sets of coplanar rays. The rays by which I see the top of a distant building may lie in the same plane as those by which I see a straight scratch on my window pane. These edges make up segments of a single line of Reid's visual geometry. So our geometric questions no longer concern things, particularly, but rather what is to be associated with the pencil of lines and planes through the eye when we interpret them as perceived points and lines. The geometry in question is the geometry of these.

Now Reid lays down the principle that the eye cannot see depth. Visual geometry is the geometry of a two-space. Let us accept this principle for now, at least. Which two-space is the one we need, though? So far we have only the pencil of rays and planes through the eye. Reid suggests that if the eye cannot perceive depth then visual space can be accurately represented by the projection of these rays and planes onto a surface which is equidistant, at every point, from the eye. So failure to perceive in depth is simulated by projection onto a surface everywhere the same in its depth from the eye. This will provide a model in E_3 of visual space. The surface in question must be a sphere, of course.

It now follows, quite simply, that every line or plane through the eye intersects the sphere in a point or a great circle. So the geometry of visual space can be properly modelled by the geometry of points and great circles on the sphere. Thus every line[2] of visual space, if followed continuously by the eye, returns upon itself. (Recall that this does not mean following the linear edge of an object extended in E_3. In an E_3 description, it means rotating the eye through the complete 360° angle of some plane of projection that passes through the eye.) For every line there is a point[3] which is equidistant from every point on the line. (Consider Angell's example of the horizon and a point immediately above the eye.) Every line intersects twice with every other line so as to enclose a space. (Consider two railway lines which seem to converge in either direction, but yet look straight.) And so on (Angell (1974)).

It is important to grasp, what Reid insists on, that the sphere represents visual space. The sphere, so far, is a surface curved in E_3; it is one

---

[1] In fact, Reid notes this application to his geometry of the wider point that in projective spaces things will look Euclidean, (1847), pp. 148–9.

[2] The reader is reminded that 'line' means 'geodesic' or 'straight (but necessarily Euclideanly straight) line'.

[3] There are two such points, in fact.

space embedded in another and with a certain posture in it.[1] But visual geometry cannot be the geometry of one space embedded in another. The idea of the eye's seeing across to the sphere works well as a device to introduce us to visual geometry, but it can be no part of the geometry proper. Visual geometry is the geometry just of a certain two-space which meets the axioms of double elliptic geometry. For vision, there is no wider manifold.

This is exhilarating and surprising. But is it correct? I fear that the answer is 'Yes and No'. What Reid wanted was an objective account of visual space, the geometry of which could be theoretically justified and intersubjectively tested (see Daniels (1972), Part 3). So we approach the problem through the geometry of perspective and the pencil of rays and planes through the eye. Only the simpler principles of projective geometry are needed here and they are correctly used. The principle of the rotating, integrating eye fits elegantly into this geometric picture. The further principle that the eye perceives no depth is deftly captured by the projective fact that a ray gives us only an equivalence class of the points along it. Nothing about the ray can tell us which of the points along it is being represented; it can say nothing about how far out the perceived point is. The visual sphere with the eye as its centre, strikes me as a remarkably penetrating way of modelling what 'no depth' perception implies. Reid gives us an illuminating and thoroughly intuitive picture of objective visual space for which I can find only applause.

But is this the only answer to questions about the geometry of vision? I do not see how we can establish that there is a unique answer. The choice between Reid's approach and others strikes me as arbitrary. I grant that it is plausible and fruitful to let the eye integrate what it follows over a period of time. But I do not see that we must adopt a principle of integration. More to the point, perhaps, I cannot see why we should pick just this principle of integration, though granting that it sheds light to pick one. Two others occur to me as equally intuitive. First, why not take visual space to include what is integrated in binocular vision? Second, why not take it to include what is integrated in movements of the eye other than the rotations by which the eye follows a line? The importance of these principles is that either of them throws doubt on the picture of visual space as two-dimensional.

The point about binocular vision is pretty direct. Whether the space

[1] In fact, there is only one posture for S_2 embedded in E_3. That is, the space cannot be bent in E_3. This need not bother us here, though.

of vision is a two-space or not is a question of topology. We can settle it by deciding what counts as a boundary for visual space. Any patch in elliptic bipolar geometry can be divided by any curve open in the patch but passing beyond it in both directions into the wider space. That is, any proper subspace of the spherical surface is topologically like any proper subspace of the plane. If Reid is correct, what we see at any moment, is just such a patch of spherical two-space. Now hold a long pencil in front of you so that it touches your nose and forehead. It lies right across the visual field. Now you have the right sort of curve (the pencil) which runs from top to bottom of the visual patch. Does this divide the space of your vision? Well, stretch out your free arm, raise a finger, and watch it as you move it across the visual field. It is always completely visible from one or other eye. So the pencil does not form a boundary in visual space and it is not two-dimensional! But is that a proof of the conclusion? Perhaps it is. What it does show, I think, is that the rules of the game for settling the dimensions of visual space are not effective. They leave all sorts of issues wide open.

What about movements other than those of following a line with the eye? A table top is placed so that as I sit my eye lies in an extension of the plane it occupies. I see the flat wooden surface just as a line – all the points in the top lie in the same plane of projection. Now I stand. The whole surface of the table comes into view. Should I integrate what I see so that, over the period of rising, a trapezoid swells out of a line in visual two-space? Or should I integrate so that I see a solid three-space figure from different vantage points? I find only arbitrary ways of settling these questions.

To get Reid's and Angell's conclusions we do need a principle of integration. I am not wanting to resist our fabricating one. But I think it is no longer a geometric question which is the right one to choose. It is a psychological question. That suggests a whole new avenue for exploration of the problem. Let us turn into it.

5 *Delicacy of vision: non-Euclidean myopia*

All of this differs from the second kind of argument I mentioned earlier which aims to explain why our visual pictures could not be grossly non-Euclidean. James Hopkins argues that the geometry things look to have is neither Euclidean nor non-Euclidean, since the differences may be so slight *in the things themselves* that they would go undetected (Hopkins

(1973), pp. 22–3). That was not my point. Hopkins also talks about the vagueness and weakness of visual imagery and how that may affect attempts to visualise on the astronomical scale (pp. 24–5). To visualise two points through which distinct lines pass, we have to imagine ourselves far enough away to see both. But then lines visualised bold enough to make an impression on us would need to be so much thickened as to merge (p. 33). I think that Hopkins is wrong about this, as I will argue in the next paragraph. But my present interest is just to distinguish the geometrical point by contrasting it with his perceptual one.

What about Hopkins' perceptual argument that the lines would merge if thick enough to be seen across the distance required? This is surely mistaken. Since spherical spaces are finite, the distances in question need never be supposed very great. If we take a geodesic joining polar points and we do suppose both points somehow visible, then we will have *infinitely many other geodesics* through both the points to choose from. If we take the mid-point between the polars on the first geodesic then we can choose another geodesic of which the mid-point's distance from the first is limited only by the dimensions of the whole space. We could choose lines of which the angle of intersection at the polar points is as close as we please to 180°. If we suppose these two such widely separated lines thick enough to merge they would fill the whole field of vision! The significant arguments on the question seem to be all geometrical and not perceptual at all.

6 Non-geometrical determinants of vision: learning to see

However, though we have made assumptions that are geometrically quite simple, they are not very realistic physiologically. Our own retinas are not plane but hemispherical. This enlarges the angle of vision. But then how things look is not simply fixed by the geometry of retinal images. When we see a straight edge, there can be no straight line corresponding to it on the retinal hemisphere. It need not even be projected onto a great circle of the hemisphere. Yet straight edges look straight to us. So how things look is not a simple matter of how the properties of their images are projected on the retina. Of course, it is usually determined by retinal projection, but in a complex way. In evolution a plane retina has never begun to compete with a curved surface of our space. We gained thereby a wider outlook on the world. But somewhere a mechanism (presumably a cortical one) corrects retinal distortion. Our hopes of finding out much about this mechanism are not bright. But let us reflect a bit further on it, nevertheless.

Seeing shapes, detecting the world's geometry by vision, is a highly sophisticated matter; it has to be learnt and relearnt and it is subject to a number of variables. People who have been born blind and who later gain their sight find it very arduous to learn to see.[1] It is small advantage to have good tactual knowledge of the world or to have mastered the concepts of a language. Visually grasping even very simple ideas, like the difference between round and square things, and whether a shape has some corners, needs long, laborious, disheartening struggles. Vision and touch may prove hard to associate. But though colour concepts are wholly new to the blind, they are comparatively easily understood because they are simple. The visual understanding of shape goes together with tactual and sensorimotor experience of things; they influence each other.

There is evidence with a different point, that how things look depends on our general experience of them and on our beliefs as to what they are like. For example, people adapt to visually distorting spectacles. The evidence is designed to show that they not only learn to cope with their changed perceptions behaviourally, but that straight things soon come to look straight, vertical lines to look vertical, though the spectacles curve or tilt them (Rock (1966)). Immediately the distorting lenses or mirrors are removed, subjects are found to make compensatory mistakes about whether the lines they see are bent or tilted. This does suggest that the way things look to them has changed. More is involved than mere retinal fatigue. The subjects adapt especially well to these affronts to visual norms when they move about and manipulate things. The way things look does adjust to our total experience of how they are. This is quite involuntary. We cannot make a straight stick immersed in water look straight simply by convincing ourselves that it is straight. Evidently, the change has to be global over what we see and it takes time. But there is every reason to conclude, for example, that Strawson is far too optimistic in his claims about the autonomy of vision. He argues for a Kantian conclusion on the grounds that vision is manifestly Euclidean and that visual images have a normative function. If manipulation of things suggested non-Euclidean shapes for them, we should not revise our view of the way things look to us – 'if the physical techniques we use fail to produce the appropriate visual effects, so much the worse for those techniques or for our handling of them.'[2] Perhaps things might

[1] Von Senden (1960). See especially Part 4.
[2] Strawson (1966), pp. 288–9.

work that way for a short time, but not for long. Even as things are, certain sorts of habitual visual environment appear to affect the way we see things. The well known impression of height and slimness given by an array of vertical lines seems to be cultural. So does the extent to which we fall victim to a wide range of other common visual illusions. Those whose visual environments are less dominated by the rectangle than is the environment of most urban dwellers, do not so readily misjudge the figures that mislead us (Segall *et al.* (1963)).

Suppose, now, our space were markedly non-Euclidean and this emerged in our manipulation of things. For example, in a finite elliptic space our bodies might be significantly large in relation to the total volume of space. Then there is every reason to think we would find that things looked non-Euclidean, even though their retinal projections, the retina still being small, were virtually Euclidean. Suppose further, that the retina is not plane but curved so as to maximise our angle of vision. Let us go back to the problem of fig. 22. In a small spherical three-space we could expect to see pairs of geodesics quite widely separated in the centre of our field of vision but obviously converging in both sideways directions. Presumably our manipulative operations with things in this small space would cause the geodesics to look straight. We would have a close approximation to the example of fig. 22 which caused us some worry earlier. There is no reason to think the lines would not look as they are: distinct straight lines that are going to meet twice.

But these last ideas do not help us to learn how to visualise non-Euclidean spaces. The point is that we cannot voluntarily decide how we will have things look to us. Visual experience and imagination are intrinsically knit into and bound by the whole gamut of our interactions with an environment which is very close indeed to being Euclidean over the regions we operate in. There is good reason to think that a strikingly non-Euclidean environment would integrate vision to motor activity in the world and thus fetter the visual imagination in different geometric chains. But, unhappily, that is not our environment. None of this enlightens us as to how we can visualise a non-Euclidean shape for space and seize it imaginatively from inside. It merely shows us that the nature of visual geometry is the purest contingency: we owe it just to a Euclidean upbringing. But though we see it is contingent, we have not seen how to avoid it. It is very far from clear what we should suppose has actually been achieved by those mathematicians who have claimed to be able to visualise non-Euclidean spaces.

4

7 Sight, touch and topology: finite spaces

So far, we have discussed really only distances and angles. This simply
follows a main tradition. But the prospect for getting what we want is
dismal when we look at it in this way. The obvious picture geometrically
– the small plane retina – suggests a largely Euclidean outlook on
distances and angles in any fairly simple space. Physiological and psychological realism about perception suggests, instead, a large but essentially
incalculable role for learning in the development of how things look to
us. However, there are plenty more opportunities for the imagination
which the non-Euclidean geometries offer us. Let us recall the fact that
many non-Euclidean spaces are finite. If they are finite we can always
imagine them small in relation to our bodies, as I suggested a bit
earlier. In that case, the curvature will be significant in our experience.
But more important than these metrical facts is the fact that finite
spaces differ topologically from E_3. These differences really do open
doors for the imagination if only we know how to make use of them.
But we will have to consider more than can be taken in at a single
glance.

Hans Reichenbach first thought of the idea that we could enlighten
our visual imaginations by teaching them some topology. Few commentators seem to have noticed this.[1] Topology has already been quite
important to us in this book: now it will get more important still. I do
not think that we can usefully visualise non-Euclidean spaces by
visualising static non-Euclidean figures in them. But if we can think
our way into a sequence of visual pictures, no one of which needs to be
dramatically non-Euclidean, we may yet arrive at a rather striking
synthesis of them. Reichenbach pointed the way. Let us turn to what he
said and consider his ideas about the problem.

Section 12 of Chapter 1 of Hans Reichenbach's book is entitled 'Spaces
with non-Euclidean Topologies'. It is a dazzling, fascinating piece of
philosophical writing. Reichenbach takes us, step by step, through the
exercise of visualising spaces which appear very different topologically
from our own space. It combines the thrill of first-rate science fiction
with the sober, less ephemeral pleasures of instruction. It is quite clear that
our experiences, generally, would differ widely in these worlds from what
they are in ours. But Reichenbach also has a second idea. He thinks

---

[1] In his survey of the literature, Hopkins cites §§9, 10, 11 and 13 of Reichenbach
(1958). He does not mention the all-important, carefully argued, brilliantly
conceived §12.

that no description of the fundamental realities[1] can entail that the space (not one of the fundamental realities) is non-Euclidean rather than Euclidean. I will argue at length, next chapter, about what this second idea means. I will also claim that it is mistaken. But I want to put the first of his arguments forward, in my own way, in what remains of this chapter.

8 Some topological ideas: enclosures

First, what is topology? Well, as we might expect, it is the study of a certain class of properties which are invariant under the transformations of the topological group. The group is easily enough defined: it is simply all 1-1 continuous transformations of a space into itself but without including transformation of improper elements such as the line at infinity.[2] The transformations are called homeomorphisms. More vividly, if less exactly, plane topology looks into whatever properties of figures would be unchanged if they were drawn on a rubber sheet which was then distorted in arbitrary ways without cutting the sheet or joining parts of it. Clearly enough, we could mould a clay sphere into a cube without tearing the surface or shearing the stuff internally. Nor would we need to join new points of it up to other points – provided we were careful. If cube and sphere can be deformed into each other in this way, then their differences in shape do not concern topology: they are topologically equivalent. But neither of these shapes can be transformed into a torus, or into arbitrarily looped and joined pretzel shapes. These shapes are topologically distinct. Topology is clearly more general than projective geometry, though it does not (usually) transform improper elements.[3]

Lines are not preserved under topological transformations, though all their incidence properties are. So are incidence properties of points and curves. If a line is mapped into a curve then the image of each point on the line is on the image of the curve. If lines (or curves) intersect in one or more points and no others then the image curves intersect in the

---

[1] A vague phrase, of course. It is clarified at considerable length in the next chapter.

[2] But see Klein (1939), Vol. 2. For reasons given in the next chapter, it is of some importance, philosophically, to reject Klein's imaginative generalisation when we are concerned with visualisation.

[3] Topology allows complete freedom to distort, but any affine stretching or squeezing has to preserve any geodesic *as* a geodesic.

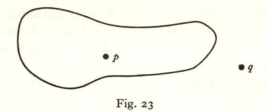

Fig. 23

images of the point or points and no others. Properties of incidence and connexion are topological invariants. We will be much concerned with the topological invariant enclosure: take a point in an area enclosed by a closed curve; then the image of the point will remain inside the area enclosed by the image of the curve. On the Euclidean plane, every closed curve encloses an area and has points inside it. This seems crashingly obvious, but exceptions in other two-spaces suggest that it will be worth a little more investigation. A closed curve (surface etc.) encloses an area (volume etc.) if and only if the area (volume) is a contractible space.[1] A space is contractible if it is deformable over itself to a point or a space of lower dimension. Thus the arbitrary closed curve of fig. 23 encloses the area of which p is a point, since we can contract the curve onto p. But the curve does not enclose an area of which q is a point since we cannot contract it onto q.[2]

In two-spaces other than the Euclidean plane, the enclosure properties of closed curves are markedly less obvious. Intuitively, it seems convincing that any closed curve on a surface must create just one contractible space. But very simple spaces topologically different from the plane lack this property. Any closed curve on the sphere creates two contractible areas which together exhaust the whole finite space. In one case we may first have to expand the closed curve to equatorial size in our 'contraction' in order to get it to contract onto one kind of point (q in fig. 24, for example) but since metrical ideas of distortion play no part in topology, the mapping still counts as a contraction. We can map the curve onto p, or onto q, so both p and q are inside it. But, on the torus, there are two classes of closed curves which do not enclose any area at all. As fig. 25 makes clear, neither of the curves a or b defines a contractible space. We can certainly map such curves homotopically onto other curves, but they never contract to a point or a space of lower dimension.

[1] Cf. Chapter 2, §6.
[2] A contraction is a kind of homotopic mapping; the intuitive picture just given will serve our purposes adequately, I think, but a dissatisfied reader may consult homotopy theory if he wishes, e.g. in Brown (1968).

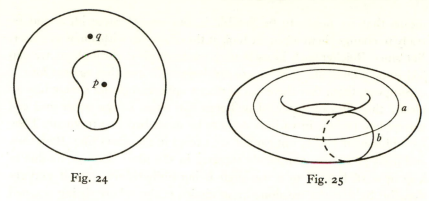

Fig. 24   Fig. 25

Thus insideness is not a metrical idea. On the sphere any point is inside all closed curves except for those that pass through the point. This is an affront to intuition and, perhaps, to common speech. We are used to thinking of insideness as a metrical idea, because, on the plane, the area enclosed by a curve is always smaller than the (infinite) area outside it. This is a mistake, since the sphere would still present embarrassing problems in the case of great circles, each of which encloses two areas of equal size: we could not settle on metrical grounds which of these was to count as the inside one.

The ideas of enclosure and contractible space are of some philosophical importance. It is pretty clear that a large number of our concepts are built on the idea that a closed surface will always contain a contractible space inside it. This has a great deal to do with the way shapes mould our concepts of common things. General terms that divide their reference often do so by suggesting a characteristically shaped surface. They consist of a particular two-space together with a definite *direction of enclosure*. The human body surface and the flesh and bones inside it, for example. This will be of some importance to us later.

9 *A warm-up exercise: the space of S_2*

With these preliminaries done, we are ready to begin the climb toward the summit of visualising the shape of eccentric spaces from inside them. Let the first stage of our journey be to imagine a complex surface as if we were two-dimensional creatures inside it. Free use of metrical concepts helps visual imagination, even though we are worried just about topology. Reichenbach thinks we can see the space in a very different way from the one made obvious by the description. This means that we ought to be flexible in our use of these ideas and be ready to change them when we look at the examples with a new eye later. But since Reichenbach himself uses metrical ideas throughout we can feel some confidence that we are not begging a question against him.

Suppose, then, that there is a perfectly spherical planet covered by an ideally thin film of water.[1] Intelligent fish swim in the water and can manipulate objects in various ways to be described in a moment. They have binocular vision, but cannot see except in the water film. Both they and light paths are completely trapped in the film: they are unable to leap upward from it or sense their being pulled down into it gravitationally; light paths are along great circles of the sphere, being trapped by the refractive index of the fluid (let us say). Let us suppose a fish's body to have a significant size with respect to the spherical area he inhabits. He has no conception of up or down but only of forwards and sideways. His conception of the world is that it has just two dimensions and that he is, himself, a creature with area but no volume.

Let us suppose the fish has a linear measuring rod. It will lie along a geodesic of the two-space (a great circle), so if he looks along its edge with one eye, it will look straight (coincide with a light path). Given two such linear rods, they will lie flush against each other however he separates, turns and rejoins them. They slide smoothly over one another.[2] This means, of course, that he must not turn the rods 'over' from a three-space point of view, else each will touch his space only tangentially at a point, bending out of the fish's two-space world. That possibility seems already covered by the provisions of the last paragraph, however. If the fish places the rod against any object which is not linear in the space, then the rod will fit on it either convexly or concavely (see fig. 26). That is, the rod may rock on the surface, or it may leave a gap when it touches. This is a simple qualitative test for how simple curves curve.

Let us suppose that there are two circular (non-linear) fences on the surface of the planet (like circles of latitude on the earth). Fig. 27 shows these fences from one elevation, perpendicular to the plane of the dotted equator, then from another elevation parallel to the equatorial plane, though one circle is hidden from this viewpoint. Each fence defines two contractible spaces: everywhere our hapless fish swims he will be inside a fence (in fact, inside both fences). What will his experience be like?

For the fish, let us suppose that both vision and progress are impeded

[1] Compare Abbott (1932).
[2] See Swinburne (1968), p. 75.

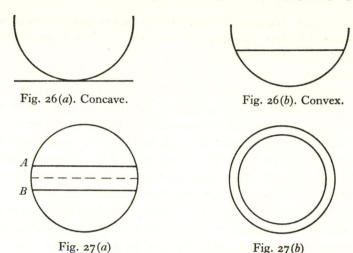

Fig. 26(a). Concave.

Fig. 26(b). Convex.

Fig. 27(a)

Fig. 27(b)

by a closed fence A. He swims toward it and finds that his rod fits against it concavely. (The fence is not a *great* circle.) He follows the fence right round with his rod, measuring its circumference. The rod touches concavely everywhere, as we would expect. He can see the fence in every direction he looks. The fence curves round behind him wherever he goes.

Suppose he now breaches the fence and swims into the region between curves A and B. From this side his rod will touch the surface A convexly: it will rock. As he follows the fence round from this side he will find it everywhere convex and curving away from him.

As the fish travels he will notice fence B which is at apparently the same distance from A everywhere. Since fence A is thin, he finds it only a little greater in circumference from the concave side than he does from the curved side.

Suppose he now goes from A to B, passing across the equatorial line dotted in fig. 27(a). He fits his rod against B and, behold: – it rocks. The fence curves convexly from this side wherever he moves along it. It is the same in circumference as fence A on *its* convex side. This could not happen with a linear rod in the two-space of the Euclidean plane. Euclidean metrical experience would suggest to us that the fence must be concave since the fish is certainly in a contractible space defined by it. Fence A would be visibly contractible from where the fish began and since fence B is everywhere at a constant distance from it on the other side from the contractible space, it must be a contractible space too. So we would expect B to curve round behind the fish instead of curving away from him. We would also certainly expect curve B to be much larger than A is. But the spherical surface has different topological and metrical features from those of the plane.

Now we let the fish break through fence B. He will find himself visibly inside it. It curves round behind him and can be seen as concave everywhere. It encloses a manifestly contractible space. The fish is still inside a fence, as he always has been, and now there is nowhere else to go.

The visual experiences of the fish will not be particularly extra-ordinary, except for some rather strange effects of movement and distance perspective.[1] The view from inside either fence will not be remarkable. It is mainly the combination of views which will be extra-ordinary. Both the shells will look convex, seen from positions between them. But as we move closer to one or the other, the appearance will change in unusual ways, a little like looking through a lens. As objects move away from the fish along a linear path they will, at first, 'get smaller' since the visual angle contracts in the ordinary way, but once they pass beyond a quadrant's distance they will apparently begin to grow in size, since the visual angle will increase. Leaving out the fences, for a moment (see fig. 28), the visual angle of a small object at p_2 seen from p_1 is plainly smaller than it is seen at p_3, though p_3 is more distant. Apparent size and distance will be much less simply related than in E_2. A similar 'lens' effect (taken twice, as it were) will give the fish an enormously enlarged view, filling the whole visual background, of the back of his own body.

10 Non-Euclidean experience: the spherical space S_3

If the model is clear up to this point then it is not very difficult to leap further and picture our experiences in spherical three-space. Just as the two-space of the sphere can be defined as a set of points equidistant from some point in a three-space, so one way to define spherical three-space, S_3, is just as the set of points which would be equidistant from a point in a containing Euclidean four-space. These higher dimensional definitions do not prevent our giving a wholly internal description of the shapes of the spaces in terms of Gaussian curvature. Nor do they forbid our visualising the internal experiences by which we could detect it. Lastly, they do not commit us to there being a real four-space in which S_3 is embedded.

[1] See Reichenbach (1958), pp. 71–3.

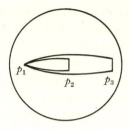

Fig. 28

The description of S_3 follows step by step the description of S_2. But now, barriers are not circular fences, but spherical shells. The enclosure relations are just as before, however.

Let us suppose that someone springs into existence in a three-space as if waking from a deep slumber. He easily sees that he is in a gravity-free spherical shell no larger than an ordinary room. It is clearly a contractible space. Beside him two measuring rods float conveniently. They always fit flush against each other however they are turned over and rematched.[1] If he sights along them they look straight. He starts to explore this new environment by measuring the shell. It is concave: each end of the rod touches a point of the shell, but the middle of the rod does not. The shell can be seen curving round behind everywhere. It is clear to vision and measurement that he is inside a spherical shell of circumference c.

Now he breaks through the shell. First he fits his rod onto the shell from this side and it rocks. It is convex, of course. He moves carefully all over it, and finds it everywhere convex and virtually the same size from outside as inside. The shell is negligibly thick. But, as he travels, he notices another shell B, everywhere the same distance from A. To be exact, if a bit laborious, for every point on A there is a point on B nearest it. The distance between a point on A and its nearest point on B is the same for all points on A. Thus shell B encloses shell A. Each of them defines a contractible space. The space was certainly contractible on the side of shell A where the explorer began. B is a shell on the other side of A: it is both a closed surface and at a constant distance from A. So B encloses A.

Now the man moves from A to B and fits his rod against the shell. It will rock! Shell B is everywhere convex, curving away from him in all directions. Its outer circumference measures just the same as that of

---

[1] Strictly, we need three surfaces, any pair of which fit flush. Then all three are plane surfaces of that space.

shell A. Yet the shells are everywhere the same distance apart. Every point on B has a nearest point on A and all these 'nearest point' pairs on the different shells are equally close. Certainly this combination of things could not happen in Euclidean three space. Now the explorer breaks through shell B and finds himself inside it. It will be slightly less in circumference from this side than from the other. It will be perfectly visible all round him, and is merely a room-sized space. The rod fits against the shell, from this side, in concave fashion. It obviously curves round behind the man at every point. So shell B encloses shell A in that both are contractible spaces on the point from which he began. But shell A also encloses shell B; both yield spaces contractible onto the point where he ends. He cannot get outside either shell, though he can break through them. There is simply nowhere 'outside' to go. The whole of the space has been traversed in a quite short time.

Vision in this three-space will be closely analogous to vision in spherical two-space as we described it before. There seems to be little point in repeating the features mentioned then. For reasons given earlier, in §3, there is no reason to think that considering perspective projection onto a retinal plane will reveal any startling new phenomenon of apparent shape. We have too few guidelines from the influence of sensorimotor activity on how things look to guess at all usefully in that direction.

11 More non-Euclidean life: the toral spaces T_2 and T_3

Having worked the trick once, let us try it again with the rather different problems set by the torus. It offers us some new opportunities since there are two kinds of closed curves on the torus which enclose nothing. No single curve of either kind can separate one region from another. The two kinds of curve were shown in fig. 25, p. 89. The series of non-intersecting curves a, b and c in fig. 29 do not enclose one another. But, taken together, they do divide the toral two-space into three distinct regions. Suppose a similar fish is trapped in a similar way on a surface with this new sort of curvature and topology. We will have some problems here when we rotate measuring rods and, perhaps, with the free mobility of the fish. The curvature of the space is not constant and things must stretch or contract to move freely. But we may suppose the rod sufficiently small not to be affected by the difference in curvature. If we inflict a few rheumatic pains on the fish, we can let him proceed

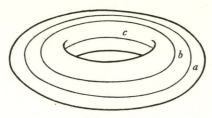

Fig. 29

much as before. We suppose the fish finds that his rod is linear tested both by light rays (which are assumed to take geodesical paths) and by being flush in every orientation in the space with another rod.

He begins somewhere between curve *a* and *b*.[1] Suppose he approaches each fence in turn and puts his rod against it. Now each of the three closed curves of the figure is geodesical on the torus, so the rod will fit flush against any of them. Nevertheless, the fish will discover that each fence has a quite definite finite length despite having no ends; further-more that, however he turns his rod to fit it against the fence, it always fits flush. Near a fence he will find that he can see right along it, so that it looks straight, though he will see his own tail in the distance.

Now suppose that he breaks through fence *b*. It will be straight from the other side also, both according to his rod and by the light ray test. Now he passes on, inside and under the curve of the torus, out of sight of the point in three-space from which fig. 29 is drawn, till he comes to curve *c*. It will have been visible to him as a complete unbroken line, none of it obscured from his view by curve *b*, which is also wholly visible to him. But we will need to look at questions of perspective more closely later. Curve *c* will also be straight both by the test of how the rod fits it and by the test of looking along it. But it will be shorter than *a*.

He breaks through fence *c*, and finds it to be straight from the other side. Fence *a* is at a distance. He crosses to it, finds it to be straight, finite yet endless and longer than either *b* or *c*. He breaches it and finds that he is back where he began. *He broke through a set of three closed, finite, boundaries and ended at his starting point*. Before we go into the topological question of boundaries in toral three-space, let us think about what the fish will *see*.

The perspective vision of the fish will clearly be very strange indeed. Given a point *A* between fences *a* and *b*, then only one geodesic through *A* fails to intersect with either *a* or *b*. That is the geodesic which closes

[1] My description differs from Reichenbach's which is careless and mistaken in various ways. See later Chapter 7.

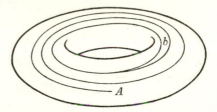

Fig. 30. 'Spiral' geodesic through A which meets b
at an apparently distant point.

back on A itself and is 'parallel' to the fences a and b. Some geodesics through A will go round the complete space many times before they meet a fence. There is no upper limit to the length of lines of sight, therefore, even though the space is finite. The longer a line of sight, the more and more tangential it becomes to the curve it meets. So from A, looking roughly toward the left of the figure into the space between the fences, the fish will see an infinitely extending line corresponding to b. Any colour variation that might be visible on b will appear to be endlessly repeated, as reflections are between a pair of roughly parallel mirrors. He will see fence a similarly. The fences will appear to approach each other as they 'recede into the distance', though a visual point separates them, provided by light following the geodesic that returns to A. But, since the fish's body occupies some area of the torus, presumably, some of this view will be blocked off by the sight of his own back end.

Now let us consider the three-space topologically like the torus. Someone is placed between two surfaces a and b (see fig. 29). Each of them looks rather flat. They appear to stretch away in all directions arbitrarily far, but with repetitions of their features visible at regular intervals. The person sees himself from behind, some distance away in all directions where he does not see either of these surfaces directly. He finds a pair of rods which look straight and fit flush however he separates, rotates and refits them. He starts to explore.

Each surface looks flat and the rod will fit flush against it. Yet the geometry of either surface is clearly spherical[1] and the measures show what the circumferences of the spheres must be. These would fit well with the visual estimates of the intervals at which marks on the surfaces appear to repeat themselves. Clearly, each one is a finite spherical shell, though also, clearly, a 'great sphere' or geodesical surface of the space he is in. That follows from its looking straight and from the flush fit of

[1] By the sum of the angles of inscribed triangles, and so on. I oversimplify a little to make the exposition run easier.

the rod. Surfaces *a* and *b*, though geodesical, have different Gaussian curvatures, different intrinsic geometries, and different circumferences. Their patterns of apparent repetition correspond to these measurements in an obvious way.

The person breaks through the spherical shell *b*. As we might expect, it looks flat (light glides along it) and proves flat by the rod test from the other side too. Measurements of this surface show that it is only a very slightly smaller geodesical sphere than from the other side: the shell is quite thin. Not far away is another surface. Vision between these two surfaces is much as it was between the others. Surface *c* looks flat and fits the rod just as the others do. Measurements on it show that it is a smaller finite geodesical sphere.

From the other side of *c*, measurements show that this surface is slightly larger than from the first side. We have passed the region which the smallest geodesic sphere would occupy. As the man moves on in the same direction, the spheres are getting larger. He finds that there is another surface not far away in the same direction. It measures almost exactly what surface *a* measured before. He breaks through this surface and finds that he is back where he began. His exploration has led him through three spherical surfaces, breaking each sphere just once. He is back where he began. Plainly, this is a non-Euclidean world.

This achieves a major objective. We began, in Chapter 1, by looking into and rejecting theories which suggest that space is not a real but a constructed entity. That led us to hope that we could explain the idea that space is an individual thing with characteristics proper to individual things. Chapter 2 prepared the ground for this by explaining how characteristics which we usually take to apply to mundane and familiar things such as hands are bound up with characteristics which belong to the space globally. This seemed to be a property of the shape of space. In Chapter 3 we aimed to attribute a definite shape to spaces quite generally. We went into some detail as to what it means to say a space has a shape (as against a posture) and how the criteria for this are ntrinsic to the space. Lastly, we translated these rather abstract concepts into sensory ones. We followed a route through strange spaces which revealed their shapes to our eyes and finger ends. The claim that space has a shape which can be imagined from inside it has some substance, and, I hope, some interest. The main problem of its justification seemed to lie simply in explaining it. That is now complete.

We have reached the goal that has been in sight over the last three chapters. But the relationists are here before us. These last results are

Reichenbach's, who went on from this point to put the whole matter in a new, thoroughly relationist, light. So far, I have pressed on with no more than passing reflection on how to set about reaching philosophical conclusions. But now we must confront issues of principle, mainly of epistemological principle, so as to keep the prize firmly in our grasp. This does not mean that we shall disagree with just one man. Reichenbach's epistemological arguments still dominate contemporary views on the question. The conclusion we have to face, either to digest or reject it, is that no real attribute of shape belongs to a space because of its global topology. Strictly, there *is* no global topology if Reichenbach is right. To say our space is S_3, T_3 or E_3 is to state no fact about it, but only to adopt a simple convention of description. The convention is, ultimately, a hollow thing. It says nothing about how the space is – certainly not how it is shaped. It reflects only what we have arbitrarily chosen, for convenience, to say.

We want to understand space as an individual thing through an imaginative grasp of it as shaped. If Reichenbach's principles of epistemology are correct then, he claims, this understanding is illusory. We must turn aside to look into just what these principles are. Then we should consider whether we ought to accept them. These are the topics of the next two chapters. Though they require a detour from the broad highway we have followed through concepts of space, it is not an incidental diversion. On the one hand, relationism has drawn most of its strength from roots that lie in epistemological soil. But on the other, epistemologies have frequently depended for their appeal on their apparent success in handling problems in our ideas about space and motion. A main interest in the philosophy of space, beyond the beauty and strength of the ideas themselves, lies in what our results show about which kinds of theories can succeed and how human knowledge can be structured in general. We can learn something more widely about our pictures of the world and how we form them.

5 The aims of conventionalism

1 Reichenbach's general strategy

The theory of philosophical method which Reichenbach defends is called conventionalism. It faces us with the sharpest danger to our conclusion so far. Many philosophers on the contemporary scene are conventionalists and certainly the great majority of those who write on the philosophy of space and time are. Conventionalism is a label covering various kindred doctrines. Some of these, like Reichenbach's, have their basis quite clearly in theories of human knowledge. Others, like Adolf Grünbaum's, do not. Yet all share an epistemological ancestry, at least. This chapter aims to give a picture from which most conventionalist doctrine differs very little. It lays out the main ideas of the theory in a largely uncritical way. There is point in a mainly expository beginning, since my disagreement with conventionalists is twofold. First, I want to argue, in Chapter 6, that conventionalism is mistaken as a theory of method. Second, I wish to go on, in Chapters 7 and 8, to argue that conventionalist methods would not yield some important results claimed for them, even if they were correct methods.

Well, then, we want to know what conventionalism is, exactly. Let us first see what it appears to be roughly. If we pick up its main outlines first, we can go into details in different areas in a more coherent, intelligible way. We want a broad impression, to start with, of how Reichenbach thinks he can deal with the examples described in §§9 and 10 of the last chapter. Let us paint his picture in a few bold brush strokes, first turning to the case of spherical space.

Suppose you are between the spheres. Your rod rocks when you place it on either of them. This seemed to be the first unequivocal sign that the space could not be Euclidean. But suppose there is a variable field of force throughout the space which, in any region, deforms all bodies in exactly the same way, no matter what the body is made of. It also draws light away from geodesical paths, so that the space will not look as it really is. The field of force deforms the measuring rod into a curve near shell *b*. Though shell *b* is really concave, it looks convex due to the warping of light and because the sharply curved rod rocks on it. If we take the rod away, rotate it and bring it back, we move it through the variably distorting field. When it is replaced, it will have been compensatingly warped so as still to rock on the shell. But any rod still looks straight since it and light are bent in just the same way. Pairs of rods can still be separated, rotated and rejoined to fit flush. Their different paths through the force field distort them so as to preserve the appearance of flatness. There will be no way whatever to detect this mysterious field of force.

You breach this shell and pass through into what looks like a finite contractible space. But, we now say, it is an infinite space, into which the distorting forces expand your body and your rod. They speed up light rays in such a way that they cross the infinite space in a quite short time. Your body becomes, in certain regions, arbitrarily distended so that you, too, may cross the infinite space. Since all substances are affected in exactly the same way by the universal forces, this infinite acceleration and swelling up are undetectable.[1]

Now it seems quite clear that the first description (§9, last chapter) and this second one are inconsistent. In the first, rods are always the same in length and always straight; in the second they change in length and straightness. In the one, both shells enclose every point which is not on them; in the other some points are enclosed by both shells, but others by neither shell. One space is finite, the other infinite. On the face of it we should conclude that one or other of the descriptions would have to be false – they cannot both be true. Or we might conclude that they simply describe different worlds. But that would be incautious and, some would say, naive. Reichenbach proposes that *both descriptions can be equally true of the very same set of facts.* He does not mean just that they are different and competing theories which explain the same evidence – the same known facts. So far as facts go, known or unknown, the descriptions are claimed *not* to differ. If we were naive, let us get sophisticated. We need to know how these two inconsistent descriptions can be equally true together.

What main features stand out in this sketch? Obviously, the two stories disagree in many ways. Equally obviously, they agree in many others. Clearly, the strategy is to persuade us that what they agree in not only is factual but also exhausts the facts. But further, the sketch is to persuade us that what they disagree in is conventional – that there would be philosophical problems if we tried to take the conflicting points as factual (as factually meaningful). The game being played is somehow or other epistemological, but we want to know a lot more about the rules.

[1] For more detail see Reichenbach (1958), pp. 77–81.

When are we justified in general in drawing a distinction between factual and non-factual expressions in language? When are we justified in doing it with a spatial language and what is the standard factual bit supposed to be in the particular case that worries us now? Let us try to get clearer by some general reflections about epistemology.

2 Epistemological basis for conventions: standard language and problem language

First, what can we say, in very general terms, about the standard factual language and about the problematic, non-factual one?

The idea that convention, not fact, makes up a great deal of what we say about the world began as part of an epistemological theory. Epistemology sometimes means the study of what men really do know and how they come to know it. We might find that we can show, in this subject, that it is not possible ever to be sure whether our space is Euclidean or not, whether the world contains any material bodies, or whether we are forever dreaming. Reichenbach did not see epistemology in that light. Nor do I. He is interested in space, not in our knowledge of it, or rather, he is interested in what spatial concepts can be made intelligible by us and what structure they have. He is not concerned with how sure we are which of these concepts apply to actual space. Conclusions about what we might (or could never) know are interesting in themselves, perhaps, but not relevant to the problem before us now.[1]

Nevertheless there is a natural and obvious picture of how people form concepts which suggests a way of coming to understand the spatial ones better. We should trace them back to their roots in the more primitive ideas by means of which we first grasped them. Presumably the more primitive ideas are all about familiar material things, since we interact most simply, directly and obviously with these. If spatial concepts puzzle us (and they do) we should restructure them somehow and the procedure of tracing them back seems the obvious way. This rather vague picture is a bit like Reichenbach's theory, but he is more sophisticated. We need to look at the scene more carefully and generally.

Suppose we start epistemology by choosing a class of statements whose truth conditions we think we grasp, clearly and directly, in our

---

[1] By contrast, Sexl (1970) is concerned just with our inability to assemble evidence which would decide between competing topological theories about the world.

simplest interactions with the world. They are to set a standard for whether other statements are clear and we will use them to explain how the other statements are to be understood. We can make later problems with complex forms of sentences a lot simpler by beginning with a set of open sentences free of truth functions or quantifiers. It is assumed that these open sentences are homogeneous in topic – an idea which I think has to stay rather inexact. But they might all be about only data of sense, for example; or all about the projective geometric properties of material things, for another example. The sentences are supposed to be true of the things (satisfied by them) in a way ideally transparent to us. Then we take the smallest class of sentences generated from these which is closed under logical operations (truth functional and quantificational). Call this class of closed sentences the *standard language*. Then it is the basis from which the remaining concepts of the language are somehow to be built.

Philosophical problems are about the world not, primarily, about language. Language is merely the obvious and tangible tool. Nevertheless, let us keep a firm grip on a linguistic, and therefore pretty definite, statement of epistemological problems. The standard language defined before is a simplified and regimented bit of natural English. Its concepts are taken to be clear. The concepts that create philosophical problems can be identified by separating out other sublanguages from English and by making them simple and formal. Which concepts cause the trouble may depend on which language we take to be standard. It may also turn out, however, that the language we select as standard depends on which concepts we think are troublesome. For example, if we pick on the data of sense as what we deal with most simply and immediately, we will find a problem on our hands about where to place material things. But if we find space puzzling we will probably convince ourselves that we ought to relate our ideas about space to our ideas about material things rather than to our ideas about data of sense. But in any case, what we need to do is to select a class of atomic open sentences, homogeneous in the topic that bothers us, and if possible, exhaustive of it as well. Then we generate from this the smallest class of sentences closed under logical operations. We can now state our epistemological problem in a pretty definite yet general way. *The problem language must be explained by the standard language*, and this means, vaguely, that we must 'construct' the former out of the latter. Putting it the other way round, we have to trace the problem language back to more primitive ideas so that we can analyse it in terms of the standard language. This still leaves a great deal vague.

In particular this is vague in saying nothing about the perceptual or, generally, mental criteria for picking a standard language.

We might see Kant's 'Copernican Revolution' in the *Critique* as the first recognition that the classical problem is impossible because it has been misconceived. In formal linguistic dress, Kant's idea is that the standard language as an input will never yield the problem language as an output unless we envisage something rather more sophisticated by way of a constructive (or analytic) program than the classical epistemologists, either empiricist or rationalist, allowed for. Kant thought he could deduce the program as a set of psychological laws simply by asking this question: what must the mind be like to make it possible that we should get the output we actually have from the input which the mind works on? Roughly, he thought that the input was the data of sense as classically conceived. Put formally, Kant's standard language was sense-datum language. One element of the output was supposed to be *a priori* knowledge of Euclidean geometry. We saw one of the results of this transcendental psychology earlier: space was argued to be a form of intuition imposed on appearance by the mind. This idea let Kant explain what he took to be our *a priori* knowledge of geometry by supposing that the program by which the mind worked on the input brought a form of outer-sense to appearances. The further details of the transcendental psychology are no longer of great interest, except historically. But the vivid flash of Kant's intellect revealed something we should not forget: a theory of the mind and its way of working is at the foundations of epistemology and *this theory is itself one of the things we have to get to know.*

Once we see this clearly, it is obvious that any psychology which aims to shape epistemology has to strike a very delicate balance: it must be modest enough to remain plausible in the face of drastic changes it may ask for in our ordinary ideas; yet it must be bold and explicit enough to give us powerful analytical tools. It is not obvious where we can turn for a theory of the mind which does both things.

3 Conventionalist commentary

Faced with this prospect modern conventionalists have been duly cautious (if not always steadily so). Perhaps we might see their methods as being justified by these reflections: Kant was surely right to see the traditional task as hopeless, but surely wrong to try to solve it by inventing a transcendental psychology. Yet some commitment to the principle of analysing and restructuring is necessary if we are to have any epistemology at all. So it will be suitably modest to begin with a standard language that deals with middle-sized, observable, material things, not speculative data of sense. But how shall we regain our old, rich, scientific theory of the world? How shall we 'construct' our problem language? The answer suggested is that we can get our theory back if we replace Kant's laws of transcendental psychology by certain conventions, which will play a similar part. They are to be seen as devices of convenience which simplify our theory and make it determinate or familiar in useful ways. So the whole language of our theory is got back again. But only some of it earns its keep by stating facts about the world; the rest holds a place only conventionally. In effect, the situation is quite like that in the most rigorous and destructive forms of empiricism. But instead of tearing the problematic fabric down, it lingers on, factually idle but somehow functional.

More explicitly, the standard language is seen as completely factual or objective. We are not concerned with who knows which standard language truths or how well any are known. The standard language is taken to be knowable in principle, and so to be conceptually in order. Language more broadly always contains this standard language. Sentences of the broader problem language will generally have some standard language entailments, but they will have other entailments as well which are not expressible in standard sentences. These other elements are factually idle and only fixed by decisions of convenience. This means that a given standard language description, which is factual and objective, can be embedded in several very different problem language descriptions. The broader, problematic descriptions will characteristically be incompatible with one another, since they are fixed by differing conventional decisions.

This is Reichenbach's position, though he sometimes wavers from it. But a conventionalist need not be mainly concerned with epistemological problems. Just so long as some argument marks out a certain segment of our theory as factual and objective, we can adopt the rest of the apparatus of conventionalist methods. For example, Adolf Grünbaum's attitude toward the problem of metrical geometries is of this kind. He has no intention to argue from an epistemological standpoint. The continuity of space is thought to show that the standard, factual language cannot be metrical. Grünbaum tells us that his ideas stem from remarks of Riemann on the problem, and these seem to be pretty purely mathematical and not at all psychological. So metrical language is fixed by some convention of congruence, and this is factually empty. But no appeal is made in Grünbaum's arguments to a theory of knowledge or of the mind.

4 Geometrical standard language: Erlanger program

But let us come back to the problem language which is before us now: it consists of all statements about global topology and whatever they may entail – about the enantiomorphy of objects, for example. Now, there is a twist in this case which we do not find in the classical task of somehow relating material objects to sensation. The standard language of sense-data is, at least arguably, untainted with ideas about physical things. But if our standard language is satisfied by material objects it is bound to contain a number of spatial concepts. We return again to the point that our concepts of material things are ineradicably spatial.[1] So we have the task of separating out an acceptable spatial language to be included *in the standard language*. But this must be quite distinct from the spatial language which contains the concepts that bother us. Until we can do this we have not articulated our problem or drawn the outline of what its solution could even look like. But how are we to make that distinction?

In Chapter 3 various metrical geometries were found to be sub-geometries of projective geometry. The situation was like this: projective geometry is the study of a certain set of projective properties P_I which are *invariant* (hence subscript 'I') properties of items in a space subjected to a certain projective group of *transformations* P_T (hence subscript 'T'). Euclidean geometry studies a group of metrical invariants E_I of a transformation group E_T, hyperbolic geometry studies the invariants H_I of the hyperbolic group H_T. The groups E_T and H_T are proper subgroups of P_T. Every transformation of E_T or of H_T is a transformation of P_T, but not vice versa. Conversely, invariants of the metrical geometries are a larger class than the projective invariants, which they obviously include. This sounds very like what we want. The standard language can have a meagre spatial vocabulary and the problem language a much richer one. We will have a beautifully sharp, theoretically sound, distinction between standard and problem language if we can show that the problem language is written in a subgeometry of the standard language. Thus, suppose we wish to claim that the problem language of metrical geometry is fixed by conventions which have been

[1] The point was made before about hands in Chapter 2.

added to a standard language which is projective. Then we can show precisely what the limits of the standard language are and precisely where and in what ways the problem language transcends it. So we can pinpoint the conventions with great precision. In fact we will be able to give a complete and exact analytical account of all the aspects of a geometrical distinction which we have pressed into philosophical service.

But how far can this ideal be generalised? Felix Klein (1939) proposed that it stretched across the board of geometries. Every geometry can be defined by specifying its group of transformations and, thereby, its invariants. Very roughly, we get a hierarchy of geometries like this:

TOPOLOGY

PROJECTIVE GEOMETRY

AFFINE GEOMETRY

METRICAL GEOMETRY (Euclidean)

in order of decreasing generality. The picture is oversimplified in various ways. We saw before that there are various types of space which differ globally in their topology, so the suggestion of a single hierarchy with a unique apex is not correct. Further, although projective transformations are roughly the group of linear topological transformations, projective geometry is not a strict subgeometry of topology. For projective geometry does not leave the line at infinity invariant, whereas topology does. Affine geometry has parallelism as an invariant but preserves neither length nor angle. It is only an intermediary stage between projective geometry and Euclidean metrical geometry, therefore.

Nevertheless, this program of Klein's (called the Erlanger program) is of great philosophical interest since it lets us separate our problems from our standards and what needs to be achieved stands out with a clarity and definiteness unequalled elsewhere in philosophy. In fact the beautiful, unrivalled precision with which the tasks are defined is a major reason why these metaphysical issues about space should attract the special interest of philosophers generally. The apparent success of conventionalist and relationalist doctrines of space has sprung largely from this kind of rigour in their foundations. In this area it is really possible to see what can be done, what cannot be done, how and why. Let us look at the Erlanger program with a more philosophical eye then. We might say that the world is factually definite only up to a topological description of things. This would tell us which subspaces are contractible, what is enclosed by what, whether things touch, interact or overlap.

But, in fact, there would be no objective question whether two objects are of the same size unless this happened to be topologically definite: for example, when both are touching along their lengths so that neither overlaps the other. The remaining metrical description of the world is wholly conventional, dictated by what we find elegant and simple to use. It is not and cannot be a matter of getting the facts right. There are no metrical facts, only metrical conventions.

5 *The special problem of topology*

Now if we do think that space is problematic, but it is also clear that we must have spatial elements in our standard language, then a great deal recommends topology as the geometrical part of the standard language. The relation of one thing's touching another, of the material points of an object forming a neighbourhood of which each is a member, of the surface of an object enclosing these points and constituting the boundary of a contractible space which encloses them, are all very tangible, immediate and familiar to us from all sorts of experience with simple manipulation of objects.[1] There are more general transformations of a space than the topological ones, but these do not give rise to an interesting set of invariants, as their name – the scattering transformations – makes clear.[2] So the topological transformations are the most general group which preserve any sort of spatial integrity for the material objects we want to start from.

But are they too general? The projective transformations preserve lines, as we saw, and this might seem a powerful recommendation. But the projective group is not really a subgroup of the topological transformations: they intersect. Topology does not 'cut or join', so it keeps objects intact. But the projective group, on the other hand, transforms the line at infinity into a finite line. More importantly now, it transforms a finite line into the improper points at infinity. Any figure, envisaged as a material object, will be torn apart under those projective transformations which map some line or plane through the figure onto infinity. If we keep these infinite elements invariant then, as we saw before, we simply get one or other metrical geometry depending on which improper elements we choose as the absolute line at infinity. That does not sound at all like a suitable basis for a metaphysical standard

[1] See Piaget & Inhelder (1956).
[2] See Courant (1964), p. 21.

language. Affine geometry does not have these disadvantages, but has never attracted much philosophical notice. Each affine space has a definite curvature, so they may seem already too problematic. Yet we shall see that there is much to recommend them (Chapter 9, §6; Chapter 10, §4, §12). Certainly, given the heavy emphasis on topological features of space in Chapters 2 and 3, it seems admirably satisfactory to select topology as the spatial part of a standard language.

The conventionalist program for metrical geometry seems beautifully clear and definite then. We take our theory of the world to be factual in its topology, but conventional in its choice of lines, parallels and metric. No topological description will ever determine the other richer description, since the metric is not phrased in terms of topological invariants. But what about Reichenbach's conventionalist program for showing that the global topology of a space is not factual but conventional? The trouble is that we do not have a more general, less explicit, spatial language which preserves the spatial integrity of things. What standard language are we to envisage if topology itself becomes soft and non-factual? Reichenbach does not put this question to the reader (perhaps not to himself) with any great clarity, yet it is obvious enough what the answer has to be. We must preserve the topology of individual objects, but the global topology of the space need not be included in the standard language. So Reichenbach claims that we can regard the global topologies of the spaces described in the preceding chapter as mere conventions. They do not enter into the factual description of the topologies of the various objects that are in the space.

6 Structure and ontology

Conventionalism, then, need not be heavily epistemological and may seem to be quite independent of problems of knowledge. It is, really, metaphysics. But it is not ontology in any very clear sense. Suppose we were to accept the data of sense as showing us what to use as a standard language for solving problems about physical things. Then one conceivable ideal, which has dominated the subject for many philosophers, is that we should completely reduce physical-thing language to the standard language. Or we might show that sentences about physical things have no factual basis but are mere conventional determinations of the standard language. In either case we get an *ontological reduction*: there are only data of sense and no distinct material things. But the present problem cannot go that way. So long as the standard language is spatial then there can be no ontological reduction of space. Only a pure language free of spatial elements, like that of Leibniz or Kant, would yield that result. What the present argument aims at is a structural reduction. If you like, it is ontological in the sense that it is about reality, not knowledge. But it excludes no *thing* from reality, only certain structures of things: metrical structures, for example. In these arguments we are concerned with the structure, not the existence, of space. That is of some importance.

7 *The basic strategy again: description and redescription*

Still, this portrays the conventionalist's problems as simpler than he sees them himself, perhaps. He does need to convince us that we should accept his language as standard and that we should agree to draw the distinction between fact and convention where he draws it. It is usually conceded that the problem language has certain advantages for us, despite its factual slack. Perhaps only it allows us to form a deterministic theory, or one familiar, coherent and comprehensible enough to liberate the imagination. Reichenbach simply says that we need a metrical geometry ((1958), p. 19) without feeling obliged to show us what we need it *for*. But, in the main, the problem language is accepted as having some kind of usefulness, or even some priority (so long as it is not a factual one) which recommends it as a vehicle for theory.

This idea gives rise to a quite characteristic conventionalist strategy: description followed by inconsistent redescription. This strategy was used in the redescription of S_3 as E_3, in the first section of this chapter. Now we can see what the description and redescription are supposed to do. Both are cast in the problem language and contain a common standard language core. So the description and redescription differ only in the problem language part. Each description is envisaged as a complete state-description in its own language, so the two problem language descriptions will conflict. This inconsistency, carefully weighed, enables the conventionalist to decide just which elements he should single out as the conventions that have globally determined the 'competing' stories in the problem language. We can then examine these to see whether it is plausible that they should be written off as conventional determinations or not.

What needs to be stressed, however, is that the problem language state descriptions count as description and redescription *only if they are identical in their factual, standard language, parts*. For example, we may give a description of some spatial region with some objects in it which assigns a Euclidean metric to the region. But then we can redescribe the region keeping our description causally deterministic by conventionally postulating forces which influence the behaviour of bodies as we move them. (Forces are regularly regarded as non-factual bits of problem language, which is a claim we will have to ruminate about later.) In this way the same facts about the coincidences among objects, their overlaps and their enclosure relations are metrically reinterpreted so as to yield, say, a hyperbolic metric for the region, giving it negative Gaussian curvature everywhere.

But these are not to be regarded as competing theories postulating somewhat remote and inaccessible facts to explain the proximate facts reported by standard language. Conventional language is not obscurely factual; it is non-factual. The description and redescription do not compete in further facts. There are none. The two descriptions are merely more or less simple, convenient or neat. What the possibility of the alternative descriptions shows is that, *if we accept them as alternative descriptions of our state of affairs*, then metrical descriptions are conventional, not factual.

8 Theory of meaning

All this very strongly suggests a theory of meaning in the offing. The suggestion is certainly true to the roots of conventionalism in logical positivism and analytical empiricism more generally. Characteristically, the problem language is regarded as not factually or objectively meaningful. Problem language disputes are written off as meaningless or as not about matters of truth and falsehood. A question is a matter of truth or falsehood if it is a matter of whether an object falls under a concept or not. Whether objects fall under concepts or not, are matters of knowledge or ignorance. None of this is meant to deny that we assign a real extension to open sentences of the problem language; that is, we select classes of objects (pairs, triples of objects etc.) and say the terms are true of them. Ordinarily, we take it that the sense or meaning of a general term or open sentence fixes its reference. Though' terms with different meanings may have the same reference we could not have terms with the same meaning but different references. Yet this is just what we do have in the case of description and redescription. The very same state of affairs is redescribed with a quite different class of ordered pairs as the extension of the problem language relation of metrical congruence. But the objects themselves, and the facts about them, do not change. So the sentences of the problem language cannot be fully, or factually, or objectively, meaningful. If the congruence classes of space are not fixed by the concept of spatial congruence we can divide the space in various arbitrary ways. It is because sense does not determine reference that convention comes to do so.

This is all a bit vague. It would be a long job to make the idea really precise but we can harden the edges of it a little by putting the matter again in terms of possible worlds. The meaning of a general term (an open sentence) is its *extension in all possible worlds*. We can get a pretty firm grip on the structural ideas at issue here by giving a set theoretical account of a range of possible worlds and of how the entities in a world fall into the sets which are the extensions of general terms. Ordinarily, we would suppose that we were dreaming up a new possible world whenever we imagine that we might include something new in the extension of a general term. Thus, imagining that Winston Churchill was among those killed in the 1940 bombing raids on London is certainly imagining a different possible world. But if a person puts forward a conventionalist thesis about some bit of the language, he is denying that this sort of change is at issue. To change the reference of problematic terms is not to envisage a different possible world. So when Reichenbach and Grünbaum envisage a different group of congruence classes of spatial intervals they are not imagining a different possible world, but only re-describing this one. When we envisage changes in the extension of standard language general terms, then we are imagining different possible worlds, which factually differ from one another. When we envisage changes in the extension of problem language general terms, then we are only noting the possibility of different conventions of redescription of this possible world.

9 Summary

Where does this leave us? So far, I have given a quite uncritical account of conventionalist argument in an attempt to make some sense of its main features, including its characteristic strategy: description and redescription are identical in their standard language components, inconsistent in their problem language components. The presentation of the two problem language descriptions is then taken to show that they are conventionally, not factually, determined. The idea that there is a standard language has its roots in epistemology. But the argument that we need to add conventional decisions to get back to the familiar and comfortable ordinary geometrical language[1] is, in a way, anti-epistemological. It springs from Kant's perception that the task as his predecessors in epistemology defined it, was impossible. But in rejecting Kant's ingenious solution through transcendental psychology, conventionalists have been inclined to cut all links with epistemology.

When we turn this battery of methodological ideas onto spatial problems we find a quite striking advantage. There is a very elegantly stratified array of languages in geometry which we can separate by appeal to the rigorous idea of groups of transformations. There seems every reason to select topology as the spatial part of standard language, given that there is no language which is free of some spatial ideas. It then turns out that the conventionalist is arguing for a reductive conclusion about the structure, not directly the ontology, of space. But if we find the global topology of space problematic, as Reichenbach does, there is no more general geometric language that will both underlie it and preserve any spatial characteristic of things. So what must be preserved is the local topology of individual things, while the global topology of space is to undergo redescription. This is Reichenbach's self-imposed task then; to redescribe the possible worlds we saw at the end of Chapter 4 so that the global topology of the world is Euclidean, but the local topology of material objects is not.

---

[1] I assume here that the language of differential geometry of general Riemann spaces as it is expressed in tensor calculus is familiar and comfortable. That is ironical of course. I have in mind that there is a standard form of geometric theory, however.

6 Against conventionalism

1 The logic of belief and appearance

We just drew a picture of conventionalism which makes the doctrine look rather rigid and formalistic. It is in the light of this picture that we will later test whether conventionalism does have the consequences claimed for it. But, you might think, I have made it look too rigid and formalistic for the kind of generality needed for a refutation of conventionalism *and kindred doctrines*. After all, you might say, the basic idea from which this bit of machinery comes is something much more natural, simple and intuitive than the picture drawn. It is the idea that the world is as it appears to be. When we go beyond saying how it appears to be, conflicts are mere conflicts in sayings.

Let me develop this idea a little further in familiar ways by recalling a familiar example. The example will be useful to us in a later chapter, too (Chapter 8). Suppose everything doubled in size overnight. How could this change ever become apparent? We can only measure things against other things, but since everything is doubled, all proportions will be as they were. The world would appear to be just as it was. The natural, simple and intuitive idea, then, is that no sense can be made of the suggestion that everything doubled in size. No such 'change' could ever become apparent. If that is the case, then it ought to be no more sensible to deny the doubling than to assert it. Nevertheless, we do deny it but wholly as a convention, in order to put senseless worries aside, perhaps.

Now, the example has its own trouble. What doubles in length, increases eight-fold in volume and mass. A shape which is stable in one size may break under its own weight if it is larger. So, as Galileo pointed out rather a long time ago, doubling in size would make itself apparent after all.[1] But troubles with the example do not directly create troubles for the principle it exemplifies. Let us look at another case where it might seem tempting to go beyond how the world appears in either of two competing ways.

[1] Galileo (1933). Of course, doubling in size might be accompanied by a change in density, to make the measures turn out the same. But this probably causes other snags to crop up. See further Schlesinger (1967) and Grünbaum (1967).

In the torus world, it seems entirely natural that the explorer should be puzzled by the fact that he sees objects and colour variations endlessly repeated. Why should he not take the world to be as it seems to be? If he simply believes his eyes might he not accept the Euclidean picture? At the same time, the toral picture also has a beautiful economy, simplicity and naturalness about it in other more theoretical ways. So it looks sensible to say that, whenever someone might formulate conflicting beliefs and theories so naturally, then we can say the difference between them is merely conventional. Perhaps the suggestion is too simple and needs some licking into shape. Maybe we will still find a central core of beliefs which the two theories share. But it does appear more flexible and fruitful to consider what a man might think and do than to follow the earlier rather inflexible and formal approach.

Sometimes Reichenbach speaks as if beliefs and theories are the core of the problem. About this very example he says 'Do we have to renounce Euclidean geometry in such a case? We do not have to, because *no one can prevent us from believing* in a pre-established harmony; *if we admit it*, Euclidean geometry is saved' (Reichenbach (1958), p. 65 [my italics]). Clark Glymour, in one of the very few discussions of convention in global topology (1972), suggests that if we were to *change our habits of reidentifying things*, we would come up with a Euclidean picture of the world in question.[1] The remarks show an inclination to move in just the direction we are thinking of.

But what principles and problems are at issue in putting beliefs to work in epistemology? There is a very simple objection to the idea that what we believe or what we say (our habits of reidentification) can decide the geometry of the world. It is a logical truism that statements of the forms:

$$\ulcorner \alpha \text{ believes } \phi \urcorner$$
$$\ulcorner \alpha \quad \text{says} \quad \phi \urcorner$$

do not entail the nested statements ϕ and are not entailed by them.[2] No one can compel me to believe that I am sitting in a chair or that I am free of pain. I might change my habits of reidentifying chairs or money, freely asserting an abundance of both. I see no reason to dispute human freedom in this. Someone might be simply obdurate, insanely credulous, muddle-headed or grossly stupid. Some of the people are some of these things some of the time. But that holds no epistemological interest.

---

[1] I am not convinced that Glymour views the suggestion with much solemnity.
[2] Except trivially, when ϕ is a theorem (or a contradiction, for the other direction of entailment).

These reflections about what I am bent on believing do not show, or begin to show, whether I am in a chair or whether I am free of pain. They provide no way of furnishing a house and there is certainly no money in them. Nor will uncertainty in belief or vacillations in what we say make the structure of space indeterminate.

This kind of point may seem somewhat familiar and elementary but it is worth looking a bit more closely at it. Though conventionalists are very sophisticated in the detail of their doctrine, they seldom look at its foundations in the contexts where they use it most damagingly. We are obliged to take our own look at these basic epistemological questions, since we want to question what is still the prevailing theory of methodology in problems about space. Modern epistemologists have focused a lot of attention on statements of the form $\ulcorner \alpha$ looks ϕ to $\beta \urcorner$. These are all expressions at least of hesitant belief, or acknowledgements of an inclination to believe the statements $\ulcorner \alpha$ is $\phi \urcorner$. But if we always place just that construction on them they are useless for any epistemological or analytical purposes. Briefly, why? First, no set of statements of the form $\ulcorner \alpha$ believes $\phi \urcorner$ begins to approach what is needed for truth conditions of ϕ. This simply repeats the truism about the logic of belief and saying. It is no less sound a point of logic for beliefs about my own pain or perception than it is for beliefs about the earth's being a sphere. Second, if we are puzzled or uneasy about the concepts in ϕ, we will be just as puzzled by those in $\ulcorner \alpha$ believes $\phi \urcorner$ since they are the same (or perhaps more complex)[1] concepts. Plainly, therefore, philosophers surely did not think that $\ulcorner \alpha$ looks ϕ to $\beta \urcorner$ was analysed along the lines of $\ulcorner \beta$ believes α is $\phi \urcorner$. The 'correct' analysans supposedly brought in simpler, purely sensory concepts; this is why they chose 'looks', 'seems' etc. rather than belief directly. Still the simpler concepts were closely enough related to those in $\ulcorner \alpha$ is $\phi \urcorner$ to do some analytical work. Obviously, the 'looks' formula was intended as a way of identifying certain belief contents. Even more obviously, the content was supposed to be sensory.

But what metaphysical machinery is supposed to justify this last idea? Does the machinery work?

---

[1] According to a Frege–Church account of belief sentences, an expression in a belief context expresses a concept of the concept it expresses outside it. I find that account an attractive one, though it does make for complexity.

2 *Belief and its bedrock objects*

Sometimes, as we say, we can hardly believe our eyes; mostly, however, we can hardly doubt them. I can do no other than believe that I hold a pen as I write these words and you cannot withold belief that you have a book before you as you read them. But it is not always like this. I may have a rather open mind about whether the rabbit was really in the hat, whether parity is preserved or who killed Cock Robin. I can give these propositions my sincere if cautious assent or my honest if hesitant denial. One motive for the procedure of tracing beliefs back to a bedrock epistemology is the vision of an ordering among beliefs as to how far it is open to us to take one or let it alone. The vision captured the profound gaze of Descartes, whom I am freely expounding. His daring idea was that we could so order beliefs that beliefs and their objects approach one another asymptotically as we proceed in the direction of greater in- voluntariness as to what we accept. In the limit, belief and its object are not one, but they are ideally intimate. In these cases no mind could be related to the object without its eliciting the belief, nor could the belief arise without the spark of its object's presence. Belief becomes irresistible but also infallible. An exciting idea, indeed.

Descartes thought he could prove at a stroke that the ideal limit which our beliefs approach lay beneath our quite involuntary beliefs about material things, including beliefs about our own bodies. The stroke was that *we might be dreaming* when we form even the most spontaneous and confident of these beliefs. Now part of this is certainly right. The state- ment that I am absolutely convinced that I now sit at a desk writing is perfectly consistent with the statement that I am asleep and dreaming. But, by itself, this does little more than repeat the truism about the logic of belief statements. The stroke becomes significant only if there is some stratum where beliefs are infallible despite the truism. Without the promise of exceptions to fallibility somewhere, the appeal to dreaming is no more interesting than any other case that exemplifies the general logical rule.

Once Descartes has brought us to this point a certain theory about the privileged stratum of beliefs starts to take shape almost by itself. That is a measure of the profound ingenuity of the method of doubt. The theory is about the operations of the mind. It has four very simple-looking and highly plausible axioms. First, dualism: the mind is a substance distinct from the body both numerically and in its 'essential nature'. It then seems plausible to say, second, that mental events and states will have mental causes and effects. However, third, there are two exceptions to this second deterministic axiom: primitive inputs from the external physical world have mental effects but not mental causes; primitive outputs to the external world have mental causes but not mental effects. Neither exception seems at all arbitrary. The inputs are data of sense, the outputs are acts of will. It follows from data of sense appearing on the threshold of the mental, unbidden from inside it, that they are untouched by theory or interpretation when they appear. Interpretation can only follow from processing in the causal net of the mind. The fourth axiom is that the mind is perfectly transparent to itself: it cannot fail to understand its own contents completely. This provides for a most elegant theory of perceptual error. Every error results from a mistaken projection from sense-data onto the external world. This now gives us precisely what we are looking for. From axiom two, every belief must have a mental cause. Beliefs about data of sense must be caused by data of sense. Axiom four, combined with the result that data of sense are untouched by interpretation, guarantees that no mistake can arise in judgements about them. So we have the stratum of beliefs we were looking for.

Historically, then, beliefs in the form of 'looks' statements came to have a role in epistemology because they lock into the machinery of a certain theory *about the workings of the mind*. In the late twentieth century the Cartesian theory of mind is certainly an embarrassment. We might wish for a theory of knowledge with no psychology in it. But surely that is not possible. The whole point of considering knowledge is that we are studying truth and the mind: how *we* form concepts, what can be intelligible to *us*, how *we* can construct a meaningful theory. So it seems useful to make this awkward commitment explicit and see what it involves us in. We can use an analogy with computers: we need to know the input to the mind, the program which operates on it and the output which this yields. It is obvious what the output is: total theory, including the problematic bits of it. The theory of input is highly developed in Cartesian thought and promises much if we buy it. Let us look further into it.

3 Basic data and standard language

The Cartesian theory was just *a priori* guesswork, though of a distinguished kind. Why did it survive so long? It follows from axiom four of the theory that the theory itself should be perfectly obvious and transparent since it is a theory about the mind. Such a result will always discourage sceptical investigation of a theory so long as it does seem simple and plausible. It certainly has seemed so. No doubt this partly explains the long survival in epistemology of theories as to the incorrigibility of judgements about the data of sense. But it is also the case that when we seek for the data, we do appear to find them. The argument from illusion is a keystone in the arch of classical epistemology. We might sum this 'argument' up, a bit tendentiously perhaps, in a sort of conditional recipe: If something you perceive appears *F* but is *G*, then look for the thing you are aware of which *is F*; that will be a datum of sense. Among those who follow the procedure, a very high proportion have felt that it has delivered the required mentalistic goods. Aided by the recipe, introspective seeking is believed to lead inevitably to finding the mentalistic data. If I make a mistake about a straight stick half-submerged in water and say it is bent, then I do seem to be able to find a bent something. Nothing relevant in the external world that I am aware of is bent, so what is bent is a mental thing. I do not make a mistake about *it* when I am in error about the stick. There can be no doubt that the imagined success of this recipe as a thought experiment has powerfully shored up the unstable fabric of the Cartesian *a priori* theory of the mind and its beliefs.

But the input theory's systematic elegance and power provides the clearest explanation for its longevity, I think. We can show these off most adroitly in the modern dress of semantic ideas but there is no reason to think the advantages were not clearly grasped by the older epistemologists. First, we can assemble a set of open sentences which are saturated by sense-data; the sentences entail nothing beyond features of the data. Second, the open sentences exhaust the data: every feature of a sense-datum is completely described in some sentence of the set. Every feature of a datum is offered directly and transparently for the mind's inspection. So the data supply a neat, apparently well defined, perfectly homogeneous and self-contained ontology for an easily separated segment of the language; that is, the language of sense-data. In other words, our immediate sensory beliefs and the sense-objects themselves match one another ideally. There is a perfect fit between words, thoughts and (mental) things. Any epistemologist must find so neat a foundation for his subject immensely appealing.

But, after all, this detour has led us back again to the methodological apparatus of a standard language. The language is not the same one as we favoured before. The motive for accepting the apparatus of method is different, too. But apart from the formalistic modern trappings that make it tangible in outline and which let us plug it into the logic machine, the standard language theory of epistemological procedure is very like the Cartesian theory. So when Reichenbach talks about what we cannot be compelled to believe he really is talking, still, within the framework set out at first. But we might now think we made the wrong choice of a standard language. Once we begin to lean heavily on belief and the naturalness of alternative theories as important methodological guidelines to our standard language, we must be supposing that the logical truism about belief sentences does not give the last word everywhere. Belief yields only belief *unless it is somewhere infallible*. That can only be at the level of sense-data, if anywhere.

4 Epistemological program and theory of mind

If we ask how the mind processes input, the answers are few and most of them rather dark. The barest form of empiricism simply allows combinations of sense-data: we can make this suggestion clear at least in the formal mode. We can construct all and only what we can get from logical operations (disjunction, conjunction, quantification) on open sentences satisfied by sense-data. The program forbids any contribution from the mind that might be seen as a content. Rationalist theories allow the mind to add some concepts and necessary truths, but how and why remains obscure. Whether translation of the problem language into the input language is needed for success in analysis is uncertain and debatable. Luckily, we can dodge these somewhat shapeless questions.

First, we must grant what I think was Kant's point. The problem as set in classical terms is insoluble. It is not quite clear what these classical terms are, but since conventionalists agree on the point we need not labour over fine distinctions. A main feature of the classical problem is that we are to begin with sense-data. Then we assume that the mind is bound to work on this sensory input by purely logical manipulations so as to produce our familiar picture of the world. There have been disputes about purity, of course, and the mind's capacities to build concepts have remained obscure to us. But Kant, at least, thought that the debate between rationalists and empiricists was idle. The former were dogmatic – there seemed no reason to assume the necessary knowledge they freely postulated – the latter were unsuccessful. At this late date, there is no reason why we should not just agree with Kant. It was this realisation which led him to transcendental psychology. It seemed that we could explain how the mind worked on the input – we could deduce the program – simply by asking *how it is possible* for us to come to know what we plainly do know. But though it is quite certain that classical epistemology needs a rather powerful theory of mind to back its methods of attack, the transcendental psychology proved too desperately speculative to provide an acceptable epistemological basis. Clearly we must weigh the credentials of these theories of mind as bits of wholly *a priori* psychology against the credentials of theories in physics or wherever else, before we can reasonably use the former as apparatus to criticise the latter. The sad fact is that we have no suitably powerful yet plausible theory of the mind's operations to do the duty Kant hoped his transcendental psychology would perform. The predicament of epistemology has become that the theories we wish to analyse have much greater claims on our belief than any theory about how to analyse them has.

5 *Some doubts about input and bedrock*

If this view of theories about how sensory input is to be processed is dim, the view of our theory about what the input is made up of, is no brighter. The picture of the mind which sprang with such vigour from the pages of Descartes now has little to recommend it as a sober scientific theory. This is not to belittle Decartes, who was inventive and early. I suggested earlier that the neat ontology of the picture recommended it to philosophers. But these systematic reasons have nothing to do with its merits as psychology. Yet the elegance of this theory is of no use to us unless it is correct psychology. Can we modify or strengthen it?

One way of getting a scientifically respectable theory of input would be to look into what happens at the body's surface. When there are significant energy changes at the skin or in the sensory receptors, nerve endings fire. A systematic description of this would give us a theory of input. We could follow neural firing below the skin and into the nerve net. But it is surely obvious that we will not get what epistemology needs, even if neuro-physiology were in a more highly developed state than it is. It could never provide us with an ontology which could be the basis of a semantics for a primitive, analysing (standard) language. The vocabulary of neurons and dendrites is simply useless for explaining ideas like 'chair' or 'table'. This is not encouraging.

We might now think it better to begin with experimental psychology. Its theory of perception has no philosophical axe to grind. But it is clear from the work already discussed in Chapter 3 that we will find very slender support for anything like an ontology which matches a theory-free language, let alone for the hope that the objects saturate the language and are exhausted by it. The work of Rock (1966) and others[1] makes it very clear that how things look is strongly influenced by how we think they are, especially when we find out by manipulating them. But if how outer things look were a matter of how inner objects are, it would then follow that judgements about the inner objects are conceptually dependent on the ideas they are supposed to analyse. Our ways of sorting and describing sense-data spring from our prior understanding of familiar physical things. Sense-datum concepts are not more primitive, then.

But this only matches the systematic advantages of ontic neatness against a systematic disadvantage of conceptual dependence. We can attack the theory far more directly and completely if we argue that there are no sense-data of the kind in question. The argument is not that we have empirical psychological evidence against them. It is rather that the philosophical argument for them is gratuitous and its conclusion implausible. I shall make the critical points as quickly as I can.

When I see a stone, a stick (in or out of water) or anything else, I am perceiving it. The same is true if I feel or hear it. That is an activity, or a state, so it is proper to raise questions about my perceiving only as they are appropriate to activities or states. My perceiving may be careless or intense, rapid or protracted, agreeable or boring to me. But it does not follow from this that when I perceive, I have a perception; for a perception is usually understood as a thing which I am mentally related to and which can be described as if it were an object with a colour, shape, texture or pitch.[2] Descriptions of the perceptual activity or state of a person do not provide a standard language through which we can analyse the rest of theory. This is not because describing the activity is only describing *overt behaviour*: it need not be overt. It is because we can only get a suitable thing-vocabulary if we suppose that whenever a person is perceiving there is a perception (a mental object) which he contemplates. On the face of it, that is an implausible and needless thing to suppose.

---

[1] See Chapter 4, §5 above.
[2] Of course, 'having a perception' might be freely used as synonymous with 'perceiving' or, more laboriously, with 'being in a perceptual state'. That is harmless, though a bit misleading.

What might tempt us to suppose it nevertheless, is just the argument (or, rather, the recipe) from illusion which I mentioned before. But the conclusion of this argument has at least two difficulties. First, it commits us to a mentalistic entity, the datum of sense, right at the very beginning of epistemology. This is clearly to bring in a problem just where things most need to be simple and secure. Secondly, it leads to the problem of the 'veil of perception'. If what I am directly aware of *is* bent when the straight stick in water only *looks* bent, then I am not directly aware of the stick. I am aware of the stick only indirectly, by mediation of the bent whatever-it-is. The stick is veiled from me by this intervening entity. Surprisingly (as it seems in the 1970s at least) its reaching this disastrous conclusion has not generally been thought to reduce the argument from illusion to absurdity. It would be welcome to avoid these two conclusions, if we can.

One intuition which might prompt us to follow the argument from illusion is the idea that there ought to be something in common between seeing a cat and having an illusion that you see a cat, including dreaming that you see one. Well, of course, the belief that you see a cat will be common. But it is plausible to suppose there must be something else in common to explain the belief. Now, certainly there may be. But the intuition looks as if it rests on the conviction that when we are in perceptual error there must be something we got right, namely what perceiving a cat and being deluded about seeing one have in common. However, I see no reason to be convinced that we could not make a completely baseless (though not uncaused) mistake. I find it quite implausible to suppose that dreams, for example, are really sequences of sense-data or any other form of perceiving or sensation. They are simply mistaken beliefs and imaginings about what we see. They arise when our minds are not fully conscious. It is not clear to me that more elaborate explanations are required for them.

In the illusion of a straight stick looking bent in water, we can grant that something in the act is bent. The retinal image is bent, for one thing, and that plays a prominent causal role in how the stick looks. But it certainly does not follow from this kind of concession that *what we are directly aware of* is bent. We are directly aware of the stick (if of any-thing) and the stick is straight however convincingly it looks bent. But not even this is the case in some illusions, such as the Muller–Lyer diagram (p. 123). Here, one line looks longer than another. But we are not aware of anything (relevant) that is longer than anything else (relevant) and nothing of the kind plays even a causal role in the illusion. We are

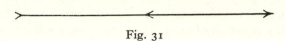

Fig. 31

simply in error. Perhaps features of our general visual environment cause the error, but that lends no support to the claim that we are aware of a something which is as the hardware looks.

This effectively knocks the bottom out of classical analytical approaches to conceptual problems through epistemology. The bottom must always be a sound theory of perceptual input. That is just what we do not have.

Our brief look at epistemology leads us to conclude that the subject needs a powerful but plausible holistic theory of psychology. The theory would have to meet systematic needs in some way rather like the Cartesian theory met them. We have no theory that comes close to doing that. But we can go on with a sort of piecemeal psychology, taking up a few isolated, but plausible, rules of thumb about knowledge. We can still say what we said in §2 of the last chapter. Our most vivid and direct contacts with the world outside our skins are with middle-sized material objects. We should still expect that to be of some importance in the structure of our concepts and knowledge of the world. It is not clear that this is a very powerful epistemological principle. After all, it would bring us back to something like the conventionalism of Reichenbach as we expounded it in Chapter 5. But let us ask, now, how conventionalism works if we try to cut it free of dubious epistemological backing.

6 Conventionalism without epistemology: some questions

We ran across some knotty problems in finding a primitive standard language which can be neatly matched with a well defined ontology. Though these problems do not block off the traditional paths of analysis and reduction, they do weaken our motives for following them. It is important to have a clear answer to the question what motive we have for the hot pursuit of analysis and reduction, in the absence of a guiding theory of the mind. Since Kant, there has been a steady drift away from emphasis on human understanding, human nature and the limits of human reason. There are two good reasons of a systematic kind why there should be some change in emphasis. Epistemology aims at rational reconstruction of our theory of the world, not at recording the actual processes of its genesis either in the scientific community or in individual learners. Further, we want to keep philosophical principles generally distinct from psychological ones, even where we are leaning heavily on psychological knowledge. But we can recognise these facts without losing sight of the part that a theory of mind and perception have to play in determining both a standard language and the procedures of analysis and reduction for any analytical epistemology. No epistemology can cut itself free from every empirical theory of the psychomechanisms of knowledge. In fact, any theory which places limits on the factual content of language in terms of limits on our powers of conception is epistemological. That is what I mean by epistemology, at least.

It does seem clear that, in some way or other, we must lower the sights from where Descartes and Kant set them. The methodological ideas of conventionalism are aptly modest, it might seem. We can replace transcendental psychology as a theory about the mental program with the idea that we impose (freely revisable) conventions on the standard language in a factually arbitrary way. There seem to be two advantages in this: we can still disdain those epistemologies which rely on speculative psychology; we can single out in one theory a class of (standard language) statements which we can prefer over the statements tainted by arbitrary conventions – *but we do not have to undergo the rigours of giving those statements up.*

A puzzle, which we must turn to later, is why a conventionalist wishes to keep these conventional, factually idle, sentences in the theory. There is the obvious advantage that we have become accustomed to them. But, if the conventionalist is right about them, it is a little surprising that they ever crept into the theory and it would seem to involve no real rigours, after all, to let them go. But let us first look at other advantages in a minimally epistemological conventionalism.

Surely we can choose a standard language according to criteria so solidly plausible as to acquit ourselves altogether of anything speculative in our psychology. We can simply take middle-sized material objects for our primary ontology and take as the standard language some familiar, rather unpuzzling set of open sentences satisfied by them. But though this leaves us clear in our minds about the bedrock in some respects, a great deal is left undecided. At what points and in which characteristic ways do conventions enter? We seem to be able to fix a central core of standard language and see it as properly factual without stretching probabilities far; but only provided we see this core as a minimum centre of the factual language. It seems unlikely that the facts run out just where we begin to get puzzled philosophically. There is no reason why the facts should not be puzzling and, perhaps, every reason why they sometimes should be. Conventions are not without their problems, then. Let us begin to look more closely at these.

Are there general features which conventions are supposed by conventionalists to share? One prominent feature which Reichenbach seems to think conventions share is that, in one way or another, they are about forces, or causes more generally. Thus he takes it that we can arrive at one among the infinite variety of metrical geometries once we postulate something sufficiently definite about universal forces, that is, those which affect all objects of whatever material in precisely the same way. We saw before that Reichenbach regards global topology as fixed provided we stipulate that there shall be no causal anomalies. We might speculate that the point of conventions of this kind is to make the total theory deterministic, especially given the importance Reichenbach takes the law of causation to have in physics. But he does not say that this is why conventions take that form in his theory, so we are just guessing. Furthermore, there is no reason to suppose that the standard language component of our theory would lapse into indeterminism if we confined ourselves to specifying causal antecedents only in that language. In any case, Grünbaum clearly rejects this whole approach and sees it only as a kind of metaphor, even for Reichenbach (Grünbaum (1964), pp. 82–3). Reichenbach is quite unequivocal that force is not a datum (Reichenbach (1958), pp. 27–8), but the idea of universal forces and the preserving of determinism evidently does not go to the heart of the problems as to what conventions are characteristically like and what they are for.

A great deal of the literature suggests that fixing *one* central convention then crystallises a whole region of the fluid problem language. Some of what Reichenbach says about coordinative definitions might easily give the impression that he is concerned just with the fact that some terms are defined ostensively.[1] This appears to have misled Eddington, for example.[2] But in later passages Reichenbach claims that geometrical forms are not data of experience. This obviously suggests that more is at work than ostensive definition. It seems that we conventionally coordinate the idea of invariance in length with some arbitrary object, dubbed as standard. This then defines equality of length (and, therefore, a geometry) wherever it is at whatever time. That is certainly not a simple case of ostensive definition. But it never gets perfectly clear what actually is at issue. Reichenbach later again speaks of the convention of

---

[1] Reichenbach (1958), pp. 14–15.
[2] Grünbaum (1964), Chapter 1, Part D.

setting universal forces at zero, which is quite unconnected with any form of ostensive definition or coordination of expressions directly with things.

Grünbaum is more simple, direct and satisfactory, in my opinion. He argues that since space is continuous, congruence (that is, length equality) must be settled in principle by the transport of some object, to be treated as a standard rod. Its length is conventionally deemed to be invariant at all times and places. Alternatively we could have conventionally deemed it to vary in length from time to time or place to place just so long as what we say preserves the factual, topological way of the world as the standard language describes it. No unusual forces or causal agencies get into the act, though which effects a given usual force may produce vary conventionally as the spatial metric varies. This is a very simple picture of Grünbaum's position. We will have to take more pains later. But it is enough to serve the present rather general purpose. It has the immense advantage of declaring itself for no more than it needs. However, it does raise the question rather sharply as to where, in general, and according to which criteria, we ought to say the language begins to idle factually and to have merely arbitrary determinations.

Before making some general objections to conventionalism, I want to distinguish it from another theory which states that truth is conventional, due to W. V. O. Quine. I think the two theories have very little in common, in fact. But Grünbaum appears to run the two together.[1] Quine regards postulation as truth by convention.[2] The prominent feature of postulates, as Quine sees it, is that we adopt them because of what they will do for the economy and deductive power of our theory of the world. We postulate on pragmatic, not evidential grounds, so what we postulate is usually significant and very general. For example the Cosmological Principle that the universe looks pretty much the same from anywhere at any time is a postulate. So is the axiom of subsets (or selection) in Zermelo–Fraenkel set theory. So again is the postulate in Special Relativity that the speed of a light signal is independent of the speed of its source. These are thought to be conventions, but not in the sense that they are not factual. Nor would they be regarded as not factual by Grünbaum. Adopting a postulate, in Quine's sense, is deciding to strike out along a certain path through the gloomy woods of theory in order that we should be going in some direction rather than in none. We stick to our arbitrarily chosen axiom even in the face of some difficulties,

---

[1] Grünbaum (1970), p. 475.
[2] 'Truth by Convention', 'Carnap and Logical Truth' in Quine (1961).

provided it continues to work pragmatically. But to call a statement a postulate in this sense says nothing at all about its meaning or factual content. Given that conventionalism seems rather heavily committed to a theory of meaning, and Quine heavily committed against any, it seems improbable that his conventions should be intended to serve the same purpose as Reichenbach's or Grünbaum's. This explains why it is that Quine regards being a postulate as a 'passing trait' of a statement. In the course of time we may feel less struck by its arbitrariness and its merely pragmatic virtues and come to regard it as solidly entrenched in evidential status. But conventionalists' conventions can hardly gain factuality as time goes on. What the facts are is decided by what the world is, not by us or by theory.

7 Some general criticisms of conventionalism

It is important to notice here a disadvantageous lack of symmetry between conventionalism and classical analytic empiricism. With the exception of some phenomenalists, philosophers have wanted to confine intelligible language within the limits of what is imaginable and meaningful; they did not wish to confine the objective world, which was conceded to be, perhaps, unimaginably rich and various. But there are two reasons for concluding that the non-epistemological conventionalists cannot make these liberal gestures toward the world. Without lapsing into the mud and ooze of a guiding theory of mind, the conventionalist can hardly construe his arguments as showing what human reason can encompass conceptually. Further, if the complexity of the world can outrun the complexity of the standard language, then extensions of language beyond the scope of the standard language can be regarded as speculative hypotheses about which facts lie beyond the pale of the preferred primitive ones. But then it would be in principle utterly unclear when we are dealing with conventions and when we are dealing with theories about the objective structure of the world. Conventionalism, then, places limits on what can be in the world and on how the world can be structured. So far as I can see, it must be among the most stringent and uncompromising reductive theories in the history of philosophy, therefore.

If this is the case, we would expect to find some powerful preliminary arguments as to why we should accept a conventionalist methodology in our philosophical approach to some subject. What we will be concluding, after all, is that various widely accepted theories characteristically misdescribe the world by overdetermining it. They give us a false picture of the structure of reality. Admittedly, the conventionalist will claim that the theory as he construes it does not really misdescribe or overdetermine; the conventional elements neither describe nor determine anything. But what is interesting in what he says is that the theory as we ordinarily take it, at face value, misleads us quite strikingly as to the objective structure of things. For example, we ordinarily think that the Empire State Building is bigger than any of us, as a matter of objective fact. We think that it remains an objective fact whether or not we are or ever were measuring our lengths along it. If this and similar things are not facts, then the conventionalist has a momentous surprise to spring on most of us. Life is full of surprises, of course, and some are momentous; I would be among the last to say that philosophy has none of them to offer. But any argument for a methodology will have to be very powerful indeed to sustain so massive a conclusion.

It is not surprising that theories sometimes underdetermine the world. *We* certainly have our limitations, even if the world does not. Coarsegrained theories can be useful even when finer-grained ones are available. It is fairly simple to use phenomenal thermodynamics, to treat gases as ideal or to regard planetary bodies as point masses. These theories are accurate enough to deal with a number of practical problems. The data on hand may be too vague or fragmentary to make any other course at all sensible. But while we recognise the practical point of simplified theories, we can surely see that they are faulty. They could mislead us by not fully mirroring the complexity of the world, though perhaps they seldom do mislead, in practice.

But there is no parallel explanation which would make sense of our using superfluously structured (hyperstructured) theories. There can be no doubt that they seriously mislead the vast majority of people who use them. If metrical geometry is a hyperstructured theory, as Reichenbach and Grünbaum claim, then almost everyone uses a too explicit theory and is thoroughly misled by it. Only the very few who follow conventionalist arguments manage to avoid being thoroughly taken in. There is no doubt that we nearly all do think that material objects have definite shapes and lengths, that objects are indeed factually, objectively, equal or unequal in length, whether they are touching or not. *If these ideas are wrong then it is of the greatest importance to rid ourselves of them, since they are widely believed without question.* That is a very simple and obvious point. There seems to be no way of enlarging on it that is very useful. But it is an important observation on conventionalism. It reveals how very strong the conclusions of its arguments are and what powerful inducements we would need to follow them. Their conclusions now look very implausible, I suggest.

I can find no real argument at all as to what the hyperstructured theory is supposed to do for us. Virtually the only comment is Reichenbach's – 'Since we need a [metrical] geometry, a decision has to be made for a definition of congruence. Although we must do so, we should never forget that we deal with an arbitrary decision that is neither true nor false'[1] (Reichenbach (1958), p. 19). But Reichenbach does not tell us what we need a metrical geometry for. Clearly he cannot say we need it for a complete account of the objective, factual world. We will certainly need one if our description of the world is to take the form we are used to, but that form is seriously misleading. If we can get used to the logic of simultaneity required by Special Relativity and to the indeterminacy principles in quantum mechanics, we ought to be able to get used to just about anything. A purely topological or projective language is already offering in a pretty thoroughly well worked-out form. I see no reason to believe that it is not deterministic within its own vocabulary. Nor can I see clear reason to complain if it is not. That would simply be how the world is, according to conventionalism. So it remains a mystery what the conventionalist thinks conventions are for.

Commonly it is said that we choose conventions on grounds of simplicity. But this is a bit ambiguous. It is easy to understand that we should choose among conventions according to how simple they are. But we are not worried about that choice right now. We are wondering *why we should choose any convention at all*, thus being lumbered with a hyperstructured theory. It is not at all obvious, to put it mildly, that there is anything simpler about having a more explicit and detailed theory. Nor would it follow on the other hand, that if we select a postulate (in the Quinean sense) because it is simple and powerful, then it cannot have a straightforward factual content.

8 Simplicity: an alleged merit of conventions

Nevertheless, there is a long-lived and plausible claim in the tradition of analytical empiricism, that the postulation of material objects simplifies any theory about the data of sense, even though the data are all that exist.

[1] The writer makes it clear that he is discussing metrical geometry.

Is the claim correct, and can we generalise from it if it is? There are two ways of taking postulates about the existence of things as mere conventions. First, the problem language is read as claiming the existence of objects over and above sense-data. That is, certain open sentences of the problem language are not satisfied by sense-data, so that other objects are 'faked up' to be values of the variables of quantification. In this case the conventionalist foresees our keeping a language which has more than merely structural excess: it has ontological overkill as well. It could hardly be clearer that the theory, as he explains it, is quite desperately misleading. But instead, the problem language may contain material objects in the same way as set theoretical language may contain virtual classes (Quine (1963), Chapter 3). The language contains devices for shortening sentences, but the expressions for virtual objects introduced in this way can be eliminated without loss in assertive power. The variables of quantification may not replace them. But now the quasi-reference to material objects really adds nothing to the standard language. The total language has greater ease of expression than the standard language without definitions, but they do not differ in any important way. Though definitions are certainly conventions, they are not the conventions that are at issue but are mere shorthand. They are only the conventions of what Grünbaum calls 'trivial semantical conventionalism' ((1964), pp. 26–7).

But whichever of the two alternatives just mentioned is the one in question, the claim that this simplifies things is fairly plausible. Perhaps, it really is easier to talk about physical things than about sense-data. That is an extremely plausible claim. The idea of a physical thing lets us unify and codify a whole complex of diverse and scattered data. But though the claim is plausible it is very far from clear that it is true. The sense-datum language which we squeeze out of the 'looks, seems, appears' vocabulary of ordinary speech is loaded with concepts whose home is in the language of material things. We do not have a fully developed, conceptually independent standard language of sense-data to compare with our ordinary language of material things. Perhaps the nearest approach is the language of Carnap's *Aufbau* (1967). Nor do we have any exact or comprehensive means of measuring how simple a theory is.[1] So it remains merely intuitive that postulating physical objects simplifies anything. It is difficult to know how much the impression of simplicity is created by the quite different fact that we are familiar with material object language and we understand it rather well.

[1] Some beginnings can be found in Goodman (1966), Chapter 3.

We would have to take this last factor of familiarity into account in estimating the alleged simplicity of metrical geometry over topology or projective geometry. The awkward fact is that we do not really understand the idea of a figure (or thing) which has only topological or projective properties and no metrical ones. Primitively, we get to understand topological or projective concepts by considering which properties of figures are invariant under certain transformations of space, where the transformations do vary other properties of the figures – including the metrical ones. The point being made is that the figure is taken to possess metrical properties both before and after its transformation. But these properties are not included in the projective language since they simply clutter it and distract attention from what is really under the microscope. Of course this doesn't even begin to prove that metrical descriptions are factual or even that they are conceptually basic. But it may make us more curious to discover what argument will have the conclusion that they can't be factual.

Whatever the case may be about the last suggestion, the parallel between an organising convention of physical objects and conventions about metric descriptions breaks down. For nothing at all makes it look probable that the structural additions of metrical geometry can play a simplifying, organising role in the way that ontological overkill might. If we are going to study metrical geometry we will also have to study the class of projective or topological invariants since they are included in the metrical invariants. It is hard to see how the hyperstructure of metrical description can be anything other than a pointless, thoroughly misleading, unnecessary, complex addition to a standard language which is perfectly adequate factually without it.

Lastly, there is the great elegance of the theory of geometric invariants under transformation to tempt us toward conventionalism when we have philosophical worries about the existence and structure of space. But the sweeping beauty with which geometry distinguishes its descriptive hierarchies ought not to lead us, by itself, into reductive paths (and perhaps it never has). No doubt it is always agreeable to operate graceful and powerful machinery, but we are concerned with truth, not pleasant exercise.

It does not look as if there can be general motives for adopting conventionalist strategy. Quite the contrary. It now seems that only some very substantial argument, rather particular in scope, could properly tempt us to a course so damaging to the structural articulation of our theory. It is a paradox, perhaps, that a theory so devoted to reducing the factual content of scientific theories should declare that it draws its inspiration and its methods from the sciences it impoverishes.

Let us leave our general picture of conventionalism, then, with a last thought in mind. When someone proves to us that two syntactically inconsistent total state descriptions have a common core, which is a total state description in some well defined sublanguage, one of two things must be true. Either the sentences in which the first two descriptions conflict are inconsistent just in *syntax*, while remaining *semantically* consistent (in this case, the conflicting sentences are conventional), or the inconsistency of the two descriptions is full semantic inconsistency. In this case the sublanguage of the common core does not exhaust the stock of primitive factual expressions, and at least one of the two competing descriptions is false. We must consider what course to adopt in such cases by a careful scrutiny of the particular area involved and of the reasons why we should discount the straightforward interpretation that we have conflicting factual claims. Our general reflections on the merits of conventionalism do not suggest that we should quickly discount the straightforward interpretation.

9 Constructing a theory

I do not think that methodological questions are of the first importance in coming to understand space in the way I think is right. But it would be unhappy to suggest, by confining my observations on methods just to critical ones, that I take conventionalists to have worried about unreal problems. It is important to make some positive suggestions about the interpretation of scientific and mathematical theories, about their construction and the constraints on them placed by what we can observe. It will be obvious that what I have to say on these questions is influenced by what Popper and Quine[1] have said about them. In common with many others, no doubt, what I think about these subjects has been prompted by their thoughts to a degree that goes beyond the power of footnotes to acknowledge. However, it would be rash to suggest that the sketch which follows would find much favour with either writer.

I favour, then, a view which sees theories as growing by bold, general conjecture and not by the cumulation of more, and more varied, evidence. The theoretician is a pragmatist and he sees his theory as a coherent

[1] Notably, in Popper (1959), (1963); Quine (1953), (1960).

whole, a growing dynamic body of theory whose development is governed by the simplicity and power of its deductive structure as much as it is fostered or constrained by what is observed.

Our window on the world is a rain of energy at the body's surface which falls hard enough to fire sensory receptors. *Nothing in the hard evidence in physiology or psychology suggests that the firing of input nerves transmits to the stage of consciousness some simple naive atom such as a sense-datum was supposed to be.* The rain of energy promotes various activities in us. Significant among them is that we join in the communal enterprise of stimulated talk. Joining this social round eventually leads us to question, answer and echo others. In this way we learn a great deal of language, theory and thought in context. Grasping how to use a sentence to respond to someone else's sentence moulds what we learn quite as much as does the more passive interaction of our receptors with the wider, but dumb, environment. Anyone's expanding grasp of a theory and a language is the flowering of a natural but socially cultivated competence. It is not the inert, inevitable assimilation of a series of separately given information atoms in the form of groups of sense-data. Once we abandon the notion that perception yields its own ontology of sense-data and accept that use in social context is a vital influence on the forms of the theory, we no longer find any clear reason for confining its primitive non-logical terms to those which fulfil some requirement of perceptual immediacy. Our metatheory of evidence for object-theoretical statements and of the growth of a theory among a community of users is, essentially, a coherence theory. We abandon the classical idea that theories grow by the accretion of basic observational atoms of information.

This kind of pragmatist theory of epistemology still insists that we structure our theory so that it is as sensitive to observation as we can make it. This means a number of things. It means that the overall deductive structure of the theory will be as simple, neat and powerful as is consistent with explaining things in general. The route from theoretical to observational sentences has to be short and clear cut. It means that a growth of epicycles in some area is a sign of its ill health, like too much fat. But this does not say much about the relation of the meaning of theoretical terms to the meaning of observational terms. A main reason for this is that it is a theoretical question what counts as an observation. This is pretty obvious if we think about what goes on in a science whose instrumental techniques are at all highly developed. That a dial is observed to read 39.456 is utterly uninformative unless we know both which bit of apparatus it is the dial of, and also the theory of what this sort of apparatus actually detects. Of course, we can dig below this conceptual level, even though our theory of physics makes sentences of the form 'Apparatus A gives reading x in the circumstance C' of primary observational importance. Though there is a gain in digging below, because we get to observations less loaded by theory, there is also a loss in the information carried by an observation report. Perhaps the loss in information carried would be worth while if we could reach a kind of observation statement which is theory-free. But there is no observation statement of that kind. Sense-data are posits just as surely as neutrinos are – in fact they are a much more dubious kind of posit, as we have seen reason to think. Material objects, of a middle-sized, observable sort, are posits, too, but surely the most universal and successful of posits. It does seem to be true that at this level we combine the minimum of theory load with the maximum of information carried. So humdrum statements about ordinary things have a good claim on our attention as observation statements. But not the only claim. Theoretically significant observations are never so simple. To overlook them is to overlook the way in which the theory of observational techniques and practical instrumental skills can influence the course of theory growth. The ideal of classical empiricism is the theory-free yet informative observation statement. But that is a Holy Grail which it now seems idle to pursue.

It follows from this that what counts as an observation or a test changes. This is nowhere clearer than in the cases which confront us most immediately in this book – measures of length and duration. But plainly, the claim that a photographic plate from a spark-chamber is an observation of a 'pi meson and a proton following the decay of a neutral lambda particle' (Overseth (1969), p. 88) or that a plate from an ion microscope is an observation of the layers of atoms at the points of a tungsten needle (Rogers (1960), p. 756) carry a high theoretical load. (This is not a remark about photographs.) Though these are far from the traditional philosopher's paradigm of an observation it would be just plain perverse to argue that the plates are not observations. But without the theory by which the complex instruments are set up to get the photographs, they are virtually nothing at all.

I urge a coherence account of observation and of theory growth, therefore. But this does not mean that I take the theory to be simply a consistent, coherent, elegant tale. What guides us in constructing it is a keen sense for its pragmatic virtues. But this does not mean that we take it to be no more than useful. Suppose we see the alternative descriptions offered by conventionalists as competing descriptions. Then there is seldom the least doubt as to which of them we will actually choose. We will pick the simplest, most 'natural', least *ad hoc* theory. There are big issues here which I raise and leave unanswered. Just what is simplicity? How is it evidence for truth? Why is it rational for us (as I believe it somehow is) to be guided by it? These are different questions from the structural ones that have occupied our attention so far. But, in any case, a theory, once held, is held as true. It is not held merely as a predictive instrument or convention. Perhaps it is regarded just as probable, but this means, again, probably *true*.

Our theory of truth is not a coherence theory. It is a correspondence theory, like Tarski's ((1956), Chapter 8). Never mind by what subtle and devious processes, in perception and linguistic social activity, the theory has grown up. The question how it portrays the world, once it is full grown, is not answered by examining the history of its growth, even when the history is rationally reconstructed. It is answered by putting the theory, as an object-language, through Tarski's mill and letting it grind out its story of how all the sentences taken to be true can be traced back recursively to syntactically (and semantically) primitive open sentences which are, simply, true of things. The things may be subtle and elusive, like neutrinos. But the relation the open sentences have to them is simple: the things just satisfy the open sentences.

10 Analysis and meaning

None of this solves the problem how to fit the *meanings* of theoretical terms somewhere into the account of the methods of philosophy. So there is still a large gap in our grasp of how to do the subject unless the whole idea of meanings or concepts would be better scrapped. This is just what Quine thinks.[1] But I am unwilling to follow this lead, though his arguments for it are impressive.[2] I think there are real problems which can be understood only as conceptual ones. For example arguments about structure are conceptual. We were in the course of one about topological structure when we broke off in order to raise these issues about method that are worrying us now. It is structure (not ontology, directly) that conventionalists want to tear down and I want to maintain. Before, I tied structure to ideas about real differences

[1] Quine (1953), Chapter 2; (1969), Chapter 1.
[2] My reasons for this may be found in Nerlich (1972), pp. 315 f.

among *possible worlds*.[1] Possible worlds are also reckoned to help us sort out our intuitive ideas about meanings. Even Quine has shown a reluctant and no doubt tepid interest in them. What do they offer us?

The meaning or sense of an expression is not the same thing as its extension or reference, as Tarski's theory gives it to us. But we can make use of possible worlds to build up a theory which connects sense and reference through the theory of truth and extension.[2] Tarski shows us how to settle the truth of every sentence in a canonical language from some not too unwieldy information about the reference or extension in the world of primitive expressions. Now we are going to say that the sense or meaning of any expression in the language is its extension or reference in all possible worlds. In any world, the Tarski truth theory gives us a comprehensive survey of the extensions of the signs, so the theory of what meanings are, and how we should handle them, becomes an elegant extension of the theory of truth.

It is vital to understand just what this theory can do and what it cannot. In one sense, meanings are (and are going to stay) primitive and unexplained. Clearly, we grasp the meaning of any general expression before we know what its extension is, or whether it is true or false (if it is a sentence). This is not to deny that 'grasping the meaning' is something we do by having some objects pointed out to us which the expression is true of. But the whole art of language is to extend the use of expressions beyond the known references by which we picked up the use. So the first thing we can do when we have learnt a language is to understand its sentences; then we have the painful task of finding out whether what we have understood is true. To put it another way: we can easily dream up a possible world; the problem is to tell whether or not the dream world is the actual world after all. A theory of possible worlds may not much improve our ability to use the language. It can only explain the ability along lines that we hope will illuminate what we do in using it and what are the general semantic principles linking the meanings with the truth.

The formal semantic model of meanings which makes them extensions of expressions in a range of worlds is a bit of machinery from the theory of sets. Any world in the model is seen as related to the actual world by a relation of accessibility. This means something like that we can 'project' an accessible world (imagine it) from the standpoint of having grasped general features (concepts) of our own world. So what carries

[1] Chapter 5, §9.
[2] See Saul Kripke's essay (among others in Linsky (1971)).

the model is our conceptual grip on actual things. The set theoretical bits and pieces simply embody this conceptual grip in a flexible way. The range of possible worlds depends somehow on the predicate stock of our current theory, so the sense of any one expression will depend on the total theory in which it is embedded. Though the range depends on what is projectible, that is something defined by the theory and a culture, rather than by the powers of individual imaginations. It may well be that no one is able to envisage whole worlds, defined in their entirety. Once the general outline of the range of worlds can be made out on the basis of our intuitive grip on imaginative projection (or on meaning) the formal machinery takes over and defines the rest in a very neat way. This leads, in turn, to a better insight into how concepts relate to one another. The suggestive power of the machinery is much greater than might be thought from this brief sketch. Finding a new conceptual possibility, like the possibility of non-Euclidean spaces, is finding access to a new range of possible worlds. When the discovery is as fundamental as a new sort of space, it puts the basic structure open to possible worlds on an entirely fresh and more general footing, so that we can expand the terms of our theory into a whole new dimension of worlds. Since the set of extensions in possible worlds is enlarged for every expression in the language, the whole theory is enriched in meaning.

Now, though the whole theory is the main determinant in how we project, observation plays some part. We could not learn the language of a theory which was in no way sensitive to observation. It is the first fact about the growth of theory about the world that we aim to maximise its sensitivity to observation together with its explanatory power. Our base for projecting other worlds is a theory made in that way. But this does not tell us much about the relationship of theory to observation in other worlds. So it doesn't tell us much about the part observation plays in the meaning of theoretical terms, either. On the one hand, we may project a world in which theory is much more accessible to observation than it is in our world. This is a common and useful technique of popular exposition. George Gamow (1939), for example, imagines a world in which the speed of light is quite slow and in which we can easily move at speeds quite near it. This strengthens our intuitive grasp on Special Relativity. But we might just as well enrich our understanding of concepts in theory by moving in other directions. For example, we will later on find it helpful to think about the consequences of General Relativity in worlds where there is no matter at all. This hardly expands

the role of observation. Both kinds of projection strengthen understanding. Classical empiricists have laid too much stress on the former,
despite the undoubted scientific importance of the latter.

But what are possible worlds themselves? Not worlds, literally. There
is only one world. 'Possible worlds' must be sets of some kind if these
ideas are to gain intellectual respectability. Perhaps they are sets of
occupied spacetime points (as Quine and Cresswell think)[1] though
I believe that is a little too simple a suggestion. We can best see what we
need from them and how sets can provide it by remembering two
purposes which we want the range of worlds to serve. Meaning is much
less a matter of pictures in the head than it is a matter of how expressions
make up well formed linguistic structures – spatially on a page or
temporally when spoken. Only certain sequences of expressions lead to
further well formed complex expressions. Possible worlds must capture
the structure of how expressions combine into meaningful complexes,
especially sentences. But the spatial (temporal) structure of sentences
touches off some causal structure in the brain (mind). I guess the causal
structure gets there by some kind of conditioning but the physio-
chemistry of this can be left obscure now. (See Devitt (1974).) What
counts is the structural features common to the spatial array (sentence)
and the causal array (nerve net).

If a Martian or a computer thinks the thought, then that is presumably a question of causal structure, too. The causal stuff may differ
widely from case to case (neurons, transistors, moveable wires and
pulleys, perhaps, on Mars), and so may the causal processes of learning.
But what is essential to the meaning is a simple isomorphism among all
these causal and sentential structures. It is just this that the set theoretical
machinery of possible worlds can capture so deftly. The causal details are
irrelevant to language: only the structures count.

We clarify meanings, then, by imagining cases – a classical technique
in philosophy. This is guided and supported by actual observation and
by theory, in ways I have sketched more boldly than precisely. So we
emerge with a whole new epistemology, mixed of much the same
elements as the familiar philosophical cake, but according to a different
recipe. We arrive at a theory by a bold and opportunistic invention
served, but not bound, by observations in various ways – by simple
sense-inspection of things, by analogies, by contextual learning and so
on. But the theory always aims at delivering observable consequences,
though it also moulds our whole idea of what an observable consequence

[1] 'Propositional Objects' in Quine (1969) and Cresswell (1972), pp. 6, 7.

is. We analyse the theory, at a primary level, by looking for its finite truth theory. We grasp the meaning of the theory (analysing it at a second level), not by building meanings exclusively out of theory-free observational bricks, but by projecting possible cases in a rather free-wheeling style, guided by our observationally sustained theory. None of this tempts us toward reduction or convention.

11 Summing up

Let me begin summing up by stressing again what these methodological questions are about. It is all too easy to see the conventionalist as arguing for the equal merits of competing theories about space (or whatever it is): one of these theories is more familiar and 'natural' than the rest, the interest in conventionalism being to see how quite bizarre theories fit the evidence just as well. But this gets conventionalism all wrong. The message is very different: beyond certain levels of structure (awfully low levels, too) there cannot really be a theory but only factually empty conventions. My aim, later on, is not to defend one theory against others, but to defend the possibility of *any theory at all at certain levels*. The first level is topology, in Chapter 7. Then I defend metrical theory as a factual thesis in Chapters 8 and 9. I ask the reader to bear in mind that the stake in these arguments is the very possibility of real topological or metrical structure of any kind. How can we find our way back to the point where we took this departure? I spent some ink on epistemological methods because we were driven into that detour by some arguments of Reichenbach. We have pursued methods farther than these arguments demanded because we foresaw some related trouble coming soon about the metric of space. That is how this picture of epistemology fits into the fabric of the book. But we might also wonder how the book fits into the fabric of epistemology in general.

Epistemology is all about the question how much man is intellectually the prisoner of his senses. It is far too easy to lose sight of this amid the often turgid and boring details of justifying a method. Can what we think leap the wall of what we see? Can the intellect freely traffic with fields, electrons, numbers – with space and time? In the spectrum of answers to these very important questions, solipsism lies at the red end. A person is utterly trapped in his private sensory world. Empiricism has generally not strayed very far from this. Conventionalism tells us that we are tied to the stake of the senses by a lead which is very much shorter than it seems to be. I have been urging us across the spectrum of opinions into the blue and toward the violet. The senses are not the captors but the guides of invention. I have argued for this especially in these last two chapters, but, really, the whole book aims to illustrate that the senses do not bind the intellect. Space is a concept both in ordinary and in highly theoretical language. It strikes me as posing the epistemological question quite crucially. I aim to show that space properly can and does figure as a real thing in our picture of the world even at the most sophisticated level. But because it is not a simply observable thing, empiricists have always aimed to sweep it under some carpet or other, to be rid of the embarrassment of giving it a prime place amid our intellectual furniture. So it is a test case and by reflecting on it at length, philosophically, I want to illustrate how reason and imagination are free.

But let us get back to problems about space itself. Even if we were to grant Reichenbach his conventionalist theory of method, would it follow that space can have no global topology?

7 Reichenbach's treatment of topology

1 The geometry of mapping S_2 onto the plane

Let us see, in this chapter, just what Reichenbach does with these conventional methods. Clearly, the strategy we talked about in Chapter 5 will lead us to discuss cases in more or less detail. But some cases are trivial and we ought to set them aside. Here is a trivial case: there is only one material thing in the universe, a wooden ball. Topologically, it is a contractible three-space with a simply connected surface. Obviously there can be a ball with this local topology in E_3, S_3 or the toral three-space. It is easy enough to run up a large number of such trivial cases, but it is not very instructive to do so. What Reichenbach needs to claim is not that it is merely sometimes possible, but possible in every case, or possible in characteristically challenging cases, to redescribe the space's global topology. We want to see how the trick can be worked in those cases where the array and the movement of objects strongly suggests some non-Euclidean global topology. In particular, it is not obvious how the apparent enclosure relations of the cases already looked at can be redescribed. A principle which Reichenbach should and does allow is that objects can move anywhere in the space[1] and, given a global topology (conventional or not) be related to each other topologically in whatever way is consistent with that space's topology.

I gave my opinion earlier that Reichenbach does not succeed in this task. However, the last two chapters have been spent on the different problem of arguing that the methodological ideas which led him to set it are mistaken. Now let us take a look at what Reichenbach actually does in his attempt to prove that global topology is conventional.

He claims that we can map the sphere onto the plane in a continuous 1–1 way with the exception of a single point. Certainly it sounds like a harmless exception. The stereographic projection of the sphere is to be the model which will direct how we can redescribe a spherical two- or three-space as Euclidean. Looking at the matter from outside, imagine that the sphere rests on the plane (or above it) and that the normal to the plane which passes through the centre of the sphere intersects the surface

---

[1] Thus Glymour scores a too easy victory, I think, in considering a spherical space in which objects never move into certain empty regions (1972).

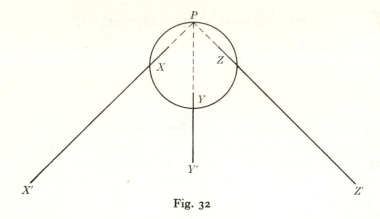

Fig. 32

at the top in a point P (as in fig. 32). This is to be the 'projection point'. Consider the pencil of lines through P. Each of them will cut the sphere in some point X (Y, Z etc.) and it will cut the plane in another point X' (Y', Z' etc.). Let X' be the image or relocation of point X in the re-description of a two-space of spherical topology as a two-space of Euclidean topology.

Now P itself is not mapped into any real (proper) point of the plane. The members of the pencil through P which intersect the sphere in P and no other point lie in a plane tangent to the sphere at P. This plane is parallel to the plane onto which we wish to project the sphere. So the mapping is 1–1 and continuous except for P itself.

Our model for a measuring rod was some arc of a great circle, when we discussed the life of a fish in spherical two-space. Any great circle which passes through P will be projected onto the plane as a line, but equal arcs of these great circles will be projected onto unequal segments of plane lines depending on how near they are to P. Equal arcs in the region 'underneath' the sphere will go into roughly equal lines of the plane, but the closer to P itself we take the arc the longer the plane line onto which it is projected. The length can increase without any upper limit. We will have to say, then, that the rod may expand or contract as it moves, but since everything which accompanies it changes proportionately (as does the speed of light etc.) this will be undetectable. Since open sentences about forces are not data, that is, not in the standard language, we can lay this limitless distortion at the door of an undetectable 'universal' force which acts on all things alike, depending only on their position. This will keep our problem language happily deterministic.

If we consider arcs on great circles which do not pass through P, then these great circles will be projected into circles of the plane, of various sizes. So, in some orientations, an apparently straight rod will be the arc of a circle; but since, for example, light will be bent round these arcs too, this will be undetectable and attributable to the same universal force as before. It is not hard to see that the redescription for the three-space example must run along closely similar lines.

2 Reichenbach's convention: avoid causal anomalies

So far, so good, then. We can regard the spherical surface as a plane at the cost of taking out only a single point. There are a few inelegant features of the redescription – undetectable forces – but these are no more than awkward. It was never part of a conventionalist's claim that each description flowed as sweetly as the other. What he said was that no facts were violated in the redescription: that is, all standard statements are retained.

But we have not yet come to what Reichenbach saw to be the crux of the matter. We want all objects – but let us fix on the rod – to be freely movable throughout the space, since they would be freely movable in spherical three-space. So the rod will move through the projection point itself. Reichenbach says that we can admit this possibility at the cost of a certain sacrifice in the problem language. There will be 'causal anomalies'. It is not obvious what that means. Universal forces are odd and plainly conventional but not causally anomalous, presumably because inventing them preserves the principle that every change is caused. I do not find Reichenbach at all clear in what he means by 'causal anomalies'. He outlines the crucial bit of the redescription in this way: we 'must admit that a material structure of previously finite dimensions can be laid down in such a way that it is situated at the same time in a finite and an infinite region without breaking'; also we 'must assume, furthermore, that a body can run through infinite Euclidean space in a finite time and return from the other side to the point of its departure' (p. 79). To allow bodies to assume infinite proportions, infinite distances, and so on is to abandon 'all previous notions of causality underlying our physical statements' (*ibid.*). Once we stipulate, conventionally, that such anomalies should be excluded from our finished theory, then we must choose spherical not Euclidean space as the space of our example. But that stipulation is a convention: causal statements including the principle of causality itself lie outside the standard language.

3 Breaking the rules: a change in the standard language

Now, it is easy to agree that something anomalous is going on, but is it causal? Let's look a bit closer. If a point a moves at a constant speed along a great circle which passes through P, then its image a' moves across the plane, ever faster as a approaches P, so as to increase its speed without limit. When a coincides with P, a' has completely traversed a half-line (ray) of the plane and does not then appear on the plane at all. But it will be mapped onto the other infinite segment of the line on the plane once a moves past P. So it will move 'out to infinity' and then 'come back from the other side'. It is quite uncertain how we ought to regard this.

But it gets clearer if we approach it more realistically by things not points. We need to look at what Reichenbach says about bodies occupying infinite regions. Suppose the measuring rod (or whatever it is) is a small, virtually rectangular, area on the sphere. The awkward spot for Reichenbach is when we move the rod so that P coincides with one of the material points of the rod. Let us agree to shed no tears over the loss of this one material point in the transformation to the plane. Our standard language is framed for material things, not material points. In fig. 33, viewed now from a point above the plane on the normal through P, the points x and y will be projected into nearby points x' and y'. But any material point of the rod will be projected beyond them, obviously. Thus, on the Euclidean plane, the rod is projected onto an object which encloses such points x' and y'. But now, in this projection, the perimeter of the rod *does not enclose any material point of the rod*, since every material point of the rod falls outside it. We have, in this projection, an infinite material object with a finite hole in it. The situation is perfectly analogous in the change from spherical three-space to E_3. The projection will work in just the same sort of fashion. Obviously, the rod has been turned inside out. Now it should be obvious what the anomaly is. *We have changed the standard language description.* The topology of the rod as a material local space is altered as the description progresses. Unless our standard description is identical in each case, we have not re-described the same facts. It will be useless to look for a yet more basic language. Local topology was just the standard language we were driven to speak in because *there is no more general fundamental spatial language than it.*

We can put what is really the same point in a number of ways. We have broken and rejoined the surface of the material area. Nor is this a

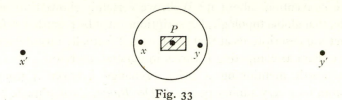

Fig. 33

trivial destruction of the integrity of the boundary. What was before a contractible material space in its local topology is now a non-contractible, infinite material space which has, really, no local topology at all except that it is a 'rod' with a local hole in it. We have breached its virgin surface and turned it inside out. We have committed the sins of cutting and joining. In breaking the intact, bounding surface, we also destroyed its enclosure of the material points. What was done is exactly what we said must not be done. Our look at the case shows that Reichenbach's redescription does not pass the test for successful conventionalism. The global topology of a space is not a conventional matter therefore. We found that we could not preserve the standard language description.

Maybe it is worth making the small, but perhaps entertaining, point that though we turned the rod inside out, we did not turn it outside in. On the sphere there are no outside points, so perhaps the suggestion that we 'put a hole in the rod' is a bit muddled. The surface of the rod does enclose that region in spherical space, though not as a larger region encloses a smaller one. But we got it quite right the first time, in saying that the redescription turns the rod inside out.

This is a general result, not one just about Reichenbach's example. For the issue at stake is whether the world can possibly have a certain degree of structure or not. It is the issue whether global topological descriptions are answered by factual structures in reality. To show that these descriptions are factually meaningless is to show they are independent of the standard language facts in all possible worlds. We can put this more clearly in the linguistic mode. However rich we make a world description in its standard language component, this will never fix the problem language description of the world's global topology. But this breaks down in the example offered. Statements about global topology are certainly factual in this description. Of course, we can think of world descriptions in standard language terms that are thin enough to leave the global topology open.[1] But are statements about global topology in such a world competing theories about the structure of the world,

[1] I pointed this out at the very beginning of the chapter.

or mere conventions about it? Well, we certainly showed, in one key example, that global topological descriptions must be regarded as factual. So structural descriptions at that level are not factually meaningless. We must see them as competing theories in general, therefore.

It is worth mentioning a brief exchange between Carnap and Grünbaum on a very similar point. In *Der Raum*, Carnap made passing mention of an argument like Reichenbach's for the transformation of the spherical earth onto an infinitely extended plane (a novel 'defence' of the flat earth hypothesis). Grünbaum suggests (Schilpp (1963), p. 670) that this means there would be a point on the earth's surface (the point of stereographic projection) which we are not physically able to cross, so that experiment would reveal it. Carnap replies (Schilpp (1963), pp. 957–8) that objects could move freely across the surface, since we only need to take out a single point. Points are not observable, so the spatial discrepancy is not observably significant.

But Carnap's reply is beside the point that is worrying us. Maybe we could never know, even in principle, whether the earth or our whole space was Euclidean or spherical or what. But we are not concerned with the problems of finding out about spaces. We are worried about the spaces themselves. Our question was whether the space was determinate in its global topology, not whether our knowledge was. Reichenbach's methods fix our attention where we want it: on space rather than on evidence.[1]

4 *Local and global: a vague distinction*

Earlier I sang the praises of the beautiful transparency of conventionalist theses when they can rest on the Klein theory of geometrical hierarchies – the Erlanger program. But there is no comparable clarity here. The distinction between local and global topologies is not at all hard. This shows, again, that we have shown something quite general about global topological structure. For example, it is surely a local feature of the topology of a space that it is contractible: we can deform it over itself onto a point. But clearly, every closed curve on the Euclidean plane creates one contractible and one non-contractible space; but on the

---

[1] An ingenious paper by Roman Sexl (1970) suffers from similar preoccupations just with evidence for competing theories. Geometrically, Sexl exploits an inversive transformation of three-space using the earth's surface as the invariant. (See Courant & Robbins (1941), pp. 140–6.) This obviously breaches local topology.

sphere every closed curve creates two contractible spaces, whereas on the torus it may create none, one or more. So which local topologies are produced by certain elements in a space depends on the global topology of the space. All we need to do is to make these elements and regions material and it looks very much as if the local topology of the material thing cannot be separated from the global topology of the space. Let us approach the same point from the other direction. I cannot see why it is not part of the local topology of a spherical shell that it encloses a certain empty region, nor can I see why it is not a perfectly straight-forward local topological relation among shells that one encloses another. We can cash this in terms of the simple, intuitive idea of a contractible space: a space deformable over itself onto a point. If both shells are contractible onto the same point, then one encloses the other. But then it follows that its being a space of global spherical topology is directly entailed by the mutual enclosure of two shells, each relation being definable by local topology. If we look at the matter in this way, Reichenbach's argument never gets off the ground. I see no reason whatever for concluding that this is a reproach to the way we chose our standard language rather than for concluding that Reichenbach was pretty obviously wrong.

Very similar arguments would show that the transformation from toral three-space to E_3 must fail. Once again, whether a closed curve or surface has the local topology of a contractible space or fails to have it may depend on the global topology of the space and how the curve is embedded in it. The topology of the spherical shells in the torus example is that of non-contractible spaces. Each shell can be homo-topically mapped (i.e. distorted) onto either of the others. But since none of them can be contracted *to a point* there are no relations of enclosure among them. Arguably, this is not local, but global. However Reichenbach's Euclidean description makes each shell a contractible space and there are certainly relations of enclosure among them. That is certainly local topology. If you like, local topology is not destroyed in the transformation, but created. That is just as bad. It means, just as surely, that the standard language element in the two descriptions is not the same.

5 *A second try: the torus*

But let us see what Reichenbach himself says about the example of the torus. The Euclidean description is taken to demand that each spherical shell encloses and is enclosed by another. Reichenbach's explorer concludes that there are not just three shells, but an infinite series of shells inside others. The curious feature of his ability to travel arbitrarily far into the nested shells is explained by supposing that there is a force contracting him and his measuring rod. Consequently, each shell is relatively larger than he is just as is required to fit the part of the toral description about how many times the rod was laid down on the surface. The explorer also comes to believe that there is massive reduplication in the world, since he is forever coming across indistinguishable scenes and objects. He writes his own name on a wall and finds an exactly similar inscription a few shells inward – also a few shells outward. He concludes that there must be 'another' explorer, indistinguishable from himself, three shells inward – and outward. Also others six, nine, twelve, etc. shells inward – and outward. All of them are doing exactly the same thing at the same time. Reichenbach sees this as another causal anomaly: there is an infinitely repeating pre-established harmony in the events and things of this world.

But why is that causally anomalous? Reichenbach asserts that pre-established harmony is inconsistent with the law of causality, which he regards as one of the most important laws of physics (Reichenbach (1958), p. 65). He takes it that the interdependence of the periodically reduplicated events cannot be regarded as ordinary causality, since it is transferred instantaneously, whereas causal influences spread continuously. Suppose that the shells in the space are semi-transparent so that the explorer can see his other self (as the Euclidean redescription demands). Then whenever he moves his arm, he observes his twin to move an arm in exactly the same way and there is no causal connexion between them.

But why should we think in this example, that one explorer's action ought to cause the other's? It seems a quite gratuitous assumption. Each man's action will be caused by neighbouring events with continuously spreading influences and the causes of the distinct actions will, themselves, be distinct. Causality is, in fact, rigorously preserved. Reichenbach claims that this is 'merely a difference in interpretation' since it requires a massive coincidence in initial conditions. This leaves one rather gasping. It is very hard to know what construction to put upon it. If causality is preserved by assuming zero probability initial conditions, then it simply is preserved. But, though I think that is a complete reply, what would follow if we accepted Reichenbach's claim? We might as well say that since the Euclidean redescription is 'merely a difference in interpretation' the space is still really toral after all. Since that would throw the thesis out from the beginning, it seems best to concede that interpretations that preserve causality really do preserve it.

But, the pre-established harmony does not seem to be the core of the matter, whether causally anomalous or not. What Reichenbach seems simply to take for granted is that a description in which there is one man and three surfaces and another description in which there are infinitely many men and surfaces are only conventionally different descriptions *of the same world*. Certainly, in the standard language as it was described before, that counts as a factual difference. It is surely overwhelmingly appealing, intuitively, to think that a world said to contain only a few things and one containing infinitely many of them are just plain different worlds, *in fact*. It seems, once again, that where Reichenbach speaks of a causal anomaly, there is really a breach of the rule that the standard language must not change. The only way in which a definite question can be posed for epistemological analysis is by preserving that rule. So we will have to see Reichenbach as having failed. Indeed, our analysis shows unequivocally that the standard language component of the description of the toral space forbids any Euclidean redescription preserving the identity of objects, let alone their local topology. Any redescription demands the infinite multiplication of things.

But there are other and sharper objections to the conventionalist's case. Reichenbach's redescription of the torus world does not give an accurate picture of the geometrical problem he faces with it. For example, he ignores the difficulty that there is reduplication not only 'inwards' but sideways as well.[1] It is clear from the discussion of perspective in the torus world (pp. 95–6, end of Chapter 4) that the explorer will see himself on all sides together with various other things that apparently repeat themselves many times over in various directions. But, more seriously, Reichenbach sees his man as moving along a path whose two-space analogue is circle *b* of fig. 25 (p. 89, Chapter 4), which crosses each fence (as we described them). But just as these circular fences blow up into spherical shells for toral three-space, so any material surface across which Reichenbach's man would move in his path from shell to shell would be a sphere intersecting the three shells. The Euclidean redescription must transform this sphere into a plane intersecting all the infinitely nested shells. At least, this is what Reichenbach's

---

[1] For this point and in the preceding paragraph, I am indebted to C. A. Hooker. See also Glymour (1972), p. 201.

procedure would dictate. But, in fact, it would be more strict geometrically – and more adroit – to project all the original 'spheres' into a
series of infinite Euclidean planes, reduplicating themselves endlessly.
In short, the transformation problem is really quite different from what
Reichenbach takes it to be. We cannot solve it by the simple means of
repeating spheres and a contracting rod. It really needs two, and
certainly needs one, class of mappings of a sphere onto a plane. But that
commits it to all the objections we found earlier in the case of equating
the topology of sphere and plane. However here the problems for
Reichenbach arise with greater directness. For we do not need to go
about to prove as before that a spherical space cannot be mapped onto
the plane without local rupture of material things in it. In the present
case we are invited to redescribe a *material sphere* as a *material plane*.
It could hardly be more obvious that the local topologies of the things
just flatly differ. I take it, then, that Reichenbach's failure to show that
torus topology is conventional fails even more clearly than his attempt
to show that spherical topology is conventional.

But, once again, where does that leave Reichenbach and us? The
enterprise of conventionalism can be nourished only on the basis of
some standard language. We also said, before, that spatial concepts
pervade all our speech about objects. There is no standard language free
of spatial ideas – not even sense-datum language is. Our preoccupation
with topology rests on the fact that this theory is written in the most
spare of spatial vocabularies. We cannot show that topological predicates
are conventional determinations of a more primitive spatial language.
There is no such language. We have just seen that the shift to a merely
local standard language will not make global topological descriptions
conventional. The conventionalist can now turn only to a sense-datum
standard language. I shall not pursue him among the stagnant pools and
faded vegetation of that retreat. I am content to point out just where the
next move must take him unless he accompanies us. I say, yet again,
that there is bite in the idea that some descriptions are mere conventions
only if it is conceded that other descriptions are not. The next section
suggests that it is really this point which Reichenbach seems to have lost
sight of. In any case, if the language of local topology is not rock bottom
then there is no objective language at all to which we can reduce spatial
concepts and in which we can analyse them.

Of course, this sheds no light on which of the competing factual
theories about global topology we should choose, but that is no longer
a question as to which sorts of structure space can have. But, in fact,

there is no doubt that we would choose the spherical and toral topologies in the cases we looked at. This is not because we found some way to check whether or not any material point is missing. We are led by the simplicity and directness of the theories chosen. This brings us sharply up against the problems whether it is reasonable to choose simple theories and why. I leave these problems untouched, save for the observation that no attempt to suck the factual structure out of our theories about space is an acceptable way to solve them.

6 A new problem: convention and dimension

There is another aspect of topology which we can best deal with at this stage. It is the aspect of dimension. Reichenbach also argued that the number of dimensions of space is conventional in just the same way as the other topological features we have been looking at.[1] We say that space has three dimensions because only that choice lets us describe the world so that the action is always by contact (or causal transmission). The convention that we ought to preserve action by contact marks out three dimensions for space uniquely. Reichenbach argues for the convention by constructing two alternative descriptions of a certain state of affairs, just as we expect a conventionalist to do, and he claims that what varies from one description to the other is just this principle of action by contact.

In the case of spherical and toral space we found Reichenbach unhelpfully vague about what had to be preserved in the competing descriptions so that both would count as descriptions of the same world or state of affairs. Here it is far more obscure what Reichenbach could have thought his standard language was made up of. Here is the case: the first description is of the (static) state at a time t of a gas of n molecules, given by $3n$ numbers. These give the position in three-space of each molecule. So we have n things and three dimensions. The second description gives a single point in a parameter space in which the point represents the state of the gas. The parameter space has $3n$ dimensions, one dimension for each coordinate of a molecule in the gas (Reichenbach (1958), pp. 275–8). So we have here only one thing and a large number of dimensions. Reichenbach says: These descriptions are evidently equivalent: either of them can always be translated into the other. In spite of this fact we consider the parameter space merely a mathematical tool

[1] Reichenbach (1958), §44. See p. 274 for example.

with no objective reference, whereas we regard the three-dimensional space as the real space. What is the justification for this distinction, which is ordinarily accepted as self-evident without further explanation?

It is not easy to persuade oneself into any puzzlement about this example. The two descriptions are obviously not equivalent in any respect but one: each characterises the state of the gas by $3n$ numbers. In the first description there are said to be n molecules, but no molecules appear anywhere in the second description. Strictly, not even the gas appears in the second description, since the point in parameter space represents only its state at t. We could not locate a gas with more or less than n molecules in this parameter space. Something far more simple and obvious than the principle of action by contact makes the two pictures different. It is only mathematically (in a rather abstract sense, too) that the pictures are alike. Are we supposed to be talking a standard language in which things never get a look in, though numbers do? If so, what reason are we offered for accepting so bizarre a choice of bedrock ideas? The example does not begin to get off the ground as providing two 'competing' descriptions with a common core of basic factual ideas.

I must admit that I am rather mystified by this section of Reichenbach's book. For he sometimes writes in it as if he thought the number of dimensions of space was not a convention at all (see the italicised passage on p. 270 especially). I can only conclude, rather reluctantly, that Reichenbach was himself tentative and puzzled in this weakest section of his brilliant book and had quite lost his grip on the principles of method that he wanted to lean on.

Perhaps he is better understood, as van Fraassen interprets him, as saying that the 'physical basis of dimensionality' is to be found in the principle of action by contact. I am not sure what the quoted phrase means and van Fraassen speaks in rather conventionalist ways in his short discussion of it.[1] But we might construe it as saying that, just as orientability is displayed in a difference between left and right hands, dimensional features are displayed in action by contact – or better, by the kind of physical thing that can be a boundary. But we have no better reason for looking on this with relationist or conventionalist eyes than we had at the end of Chapter 2. This is no more than pointing out what best displays a global topological feature to us.

[1] Van Fraassen (1970), pp. 137–8. In the italicised sentences (p. 137) he speaks of our space being 'called' the coordinate or 'real space'.

The global topology of space is an empirical fact about it. It is not a convention. We have to discover what the topology is and the task may not be easy. Let me stress that we did not need to overthrow conventionalist theories of method to get this result. The argument was that we could not show topology to be non-factual by conventionalist patterns of argument. That is because topology is so basic and general. There is nowhere to retreat from topology to some more ultimate standard language in which to probe and test topology. At this point a conventionalist's back is to the wall. But the arguments so far have done nothing to suggest that the picture will be at all like this if we think about the metric of space instead. For if we want to show the metric to be a mere convention we do have a well-defined, more basic language to test it by: the language of topology. So we are far from having finished with conventionalism, even though we have concluded that topology is safe from it.

The result that topology is safe is an important one. It means that all the main results we have reached so far can be kept. The strategy of the earlier Chapters 1–4 was to place most of the weight on topological features. Almost all the argument of Chapter 1, which first suggested to us that space is a particular, is about the likenesses and differences between the topological aspects of space and of systems of relations. Chapter 2 made free use of metrical ideas (axes and planes of symmetry, for example), but the heart of the matter – the orientability and dimensions of space – was topological and it was really the topological features that prompted us to make sense of the idea that spaces have shape. When we discussed shape and how we might make sense of it in the imagination, we raised projective and metrical issues at some length. Those still fall under the shadow of conventionalism. But in the end, we fell back on topological differences among spaces to give us our picture of the shape of space from inside. We made free use of metrical crutches for the imagination again (we had the measuring rod rock on a convex shell, for example); what we are still aiming for is a metrical understanding of the idea of the shape of a space from inside it. But 'topological shape' still leaves us with a broad picture of shape[1] which is enough to do a great deal of the metaphysical work for us. It will give us a basic understanding of what it means to say that space is included in our ontology as a particular thing with shape in a clear-cut sense. If the conventionalist is right about the metric, then not only will space fail to

---

[1] A Moebius elastic band differs in shape from the ordinary elastic band in a way which does not depend at all on how we stretch them.

have a metrical shape – so will everything else. Topological shape will be the only kind of shape, in fact. We can be confident therefore that the shape of space will be as definite as the shape of anything can be. Space is a particular with a definite structure, a topological one, just like any other particular thing. Space is not a nothing but a something. It is a real live thing.

8 Measuring space: fact or convention?

1 A new picture of conventionalism

In Chapter 6 I said that there is no general justification for seeing parts of our theories in science or mathematics as mere conventions, empty of factual meaning. A general epistemological justification is always an attempt to bind theories to limitations in our powers of conception. At any rate, that is what I mean by 'epistemology'. But any epistemological conventionalism with real bite needs a theory of mind much more powerful than any we have or are likely soon to get. Conventionalism without epistemology binds the factual meaning of theories to limitations alleged to lie in the scope of things themselves. That is, things are supposed to be incapable of the degree of structure which the theory (taken at face value) gives them. I said, in §§7–8 of Chapter 6, that though conventionalism in general is implausible we might find certain 'powerful preliminary arguments' which could apply in the case of certain of our theories. A case of this sort crops up now, when we turn to reflect on our theory of the sizes of things in space. It is said that the metric of space cannot be purely factual and must be partly conventional.

This idea has both popular and distinguished support. It is very easy to get people to agree to the epistemological argument we looked at briefly in Chapter 6. This was that it is senseless (not factual) to say that everything – simply everything – doubled in size overnight, and also senseless to deny it. This is the thin end of a wedge that splits the idea of a factual metric wide open. Among the great, Leibniz (Alexander (1956)); Riemann ((1929), pp. 413, 425); Clifford ((1955), pp. 49–50); Poincaré ((1913), Chapters 2 and 3, (1899), (1946), (1898), (1900)); Russell (1899); Einstein 'The Foundations of the General Theory of Relativity', p. 161 and 'On the Electrodynamics of Moving Bodies', pp. 38–40, in Einstein *et al.* (1923), and 'Reply to Criticisms', pp. 676–8, in Schilp (1949); Carnap (1922), and Reichenbach (1958) all argued against a factual metric. At present conventionalism's most ingenious, penetrating and persistent apologist is Adolf Grünbaum (1964), (1968), (1970). Taking his core idea from Riemann, Grünbaum has developed a theory about why the metric of space must have an ingredient of convention which our earlier reflections on method leave untouched.

The theory has been widely influential. I will focus attention on it exclusively in this chapter.

A preliminary sketch of strategies will help us to see which way the wind is blowing when we get down to argument in detail. We are looking for a new kind of explanation as to why we should take metrical theories about space as conventional. It is not an explanation concerned in any way with limitations on our powers of conception (Grünbaum (1968), p. 4). Grünbaum has to choose some new way of telling us what he means by a convention and some new way of convincing us that any metric for ordinary space must be a conventional thing. He uses a structural argument. He appeals to the make-up of intervals in discrete space to show that they can have an intrinsic metric – various intrinsic metrics, in fact. He then argues that the make-up of continuous space does not allow the kind of structure that provides intrinsic metric, so the metric must be conventional. As our earlier reflections would lead us to expect, the difference between discrete and continuous spaces as to how their intervals are made up, is topological. Clearly, what is suggested here is that the intervals themselves limit the structure they can actually have. The structure of metric theories of continuous space outruns the structure of the facts which the theories are about.

I give an outline of Grünbaum's theory in §2. The main ideas of the theory are expounded at greater length in §§3–6. This longer exposition is rather technical and more arduous to follow than most of this book since I intend it to do justice to the latest and most careful of Grünbaum's expositions (1970). The reader will follow the criticisms of Grünbaum's theory which I offer in §§7–11 if he understands §2. Some readers may wish to skip §§3–6 for this reason. Those who do so should bear in mind that the final test of whether my criticisms succeed does not lie in their plausibility as a reply to the arguments of §2, but in their dealing justly with the full-blown theory.

2 *The conventionalist theory in outline: continuous and discrete spaces*

The measurement of space tells us how the size of one interval or area or volume compares with others. We need to know which regions equal which. Among the unequal ones, we need to know which are the larger regions and by how much they are larger. If the regions coincide, the task is easy: they are equal. But that only allows us to decide the empty question whether each region is exactly as large as itself. Regions of space do not move: a region coincides only with itself. Evidently, we have not made much progress if we can decide only that each region is exactly the size it is. Less trivially, if one region contains another, we can be sure that it is larger than the one it contains. But we are not yet able to settle by how much it is larger.

But how can we compare two regions that overlap without either including the other? How are we to compare two regions that have no common point at all? The suggestion Riemann makes is this: 'Measurement consists of *superposition* of the magnitudes to be compared; for measurement there is requisite some means of *carrying forward* one magnitude as a measure for the other' ((1929), §1.1, p. 413 [my italics]). Since spatial regions are immovable, we cannot carry forward one region itself so as to superpose it on distinct but equal regions. We will have to carry forward something else: a rod is the obvious thing to carry. It is in space, but not part of it, and it will give us a familiar basis for saying when distinct regions are equal and by how much some exceed others in size.

This last technique of measurement can never be intrinsic to any pair of regions we use it to compare, nor is it intrinsic to the space as a whole. This idea of not being intrinsic is important. It means that we cannot compare the regions without there being something distinct from them both (and from any part of both) which is our means of carrying a magnitude forward. Nor is there some other purely spatial thing apart from the regions to carry about. By contrast, if one region contains another then we can regard the first as intrinsically greater than the second. For here, there is something about the structure just of the regions themselves which makes the one larger than the other. In the former case, there is nothing about the structure of non-overlapping regions which can settle their relative size. It is not that we cannot find out their relative size without a rod. It is that, in fact, the one is neither intrinsically larger, smaller nor the same in size as the other. There never get to be facts of any kind about their relative size, until after we decide how to relate them extrinsically. Riemann is not making an epistemological point, but a structural one. So Grünbaum interprets him, at least.[1]

He contrasts this state of affairs with how things would be if space were not as we usually think it is. We think space is a continuum and

---

[1] I have some misgivings about this interpretation, and even stronger ones about a parallel interpretation of Weyl. See Grünbaum (1963), §2c. i; Riemann (1929), §§1 and 3; and Weyl (1922), §12.

this means, among other things, that we can divide any region up, endlessly, into smaller and smaller subregions. But suppose instead that we came to an end somewhere in this process of division: suppose that every region is composed of discrete spatial chunks, atoms, grains or quanta which have no spatial parts, though they are extended.[1] I will call them 'grains' from here on. Now we can compare regions by counting up the number of grains in each and seeing which has the more in it. We could say that regions are equal if they have the same number of grains, that one is larger than another if it has more grains and that the ratio of the sizes of regions is given by the ratio among the numbers of grains they contain. The point is not that we would be able to tell which is which in this way. It might be no more feasible in practice to count grains that to count molecules. We might still rely on rods, perhaps in the context of some theory about rods and grains. The point is that what makes the metric is the granular structure of regions. In this supposed example, the metric is intrinsic: we appeal only to regions and how they are made up.

We cannot appeal to a similar structure if space is continuous, or even if it is just infinitely divisible and dense.[2] But what lies behind this claim is not merely the distinction between what is intrinsic and what is extrinsic. We need to find something in continuous space that is both intrinsic *and adequate* for capturing what we intend by the idea of a metric. If space is dense, then between any two points there are always others and any region, or even any interval, has at least a countably infinite set of points in it. If the structure of space were just that it is dense and countable, then every region or interval would have exactly as many points as every other: as sets of points they would all have the same cardinal number, $\aleph_0$ (aleph null) and they would be ordered as the rational numbers. Clearly, this would not give us anything like as adequate a basis for the metric as we get in granular space, since a dense region has no more points in it than any of its subregions. Given a region R_1 and a subregion R_2, consider a third non-overlapping region R_3. It will contain just the same number of points as the first region R_1 and as its subregion R_2. So the point-structure of regions gives us no adequate basis for equating the non-overlapping region R_3 with the subregion R_2 rather than the containing one R_1. We will be no better off

---

[1] The distinction between discrete and continuous spaces is a topological one. See Brown (1968), Chapters 1 and 2, especially Chapter 2, §5.

[2] The rational numbers, in their usual ordering, are dense. However, the rational numbers are not continuous, though the real numbers are.

if the space is continuous, each interval having an uncountable infinity of points, with irrational as well as rational points, having the order-type of the real number interval. The number of points in any interval or region is still the same as in every other, only now the number is that of an uncountably infinite set. Nothing about the point-structure of regions makes up an adequate metrical basis, so we will need to find the basis extrinsically, by carrying a magnitude forward. Roughly,[1] any region or interval in space is 'made up' out of its points in the same way as any other.

Now, if there is no intrinsic and adequate metric for a space, Grünbaum describes it as 'metrically amorphous'. If we give it an adequate metric, despite this amorphousness, then we must do so by adopting a convention about the means for carrying a magnitude forward. That does seem very plausible indeed in the light of what has been said so far, since there are obviously several things we might do which could be regarded as transporting a magnitude. We can transport a rod, but we can also use the wavelength of some standard kind of oscillator. What we choose and how we use it looks quite arbitrary. We can hardly suppose that a rod, or whatever it may be, has an intrinsic length if intervals themselves do not. That a rod has length is simply a matter of its being extended through some interval or region.

It is interesting that Grünbaum now thinks that discrete space can have various metrics each of which is both intrinsic and adequate (Grünbaum (1970), §3). He gives several examples. One (called M_1) seems to be the obvious case where each grain (but strictly, the *measure of the unit set* of each grain) is assigned the same value. Another (M_2) gives each grain a slightly different size from every other; yet a third (we could call it M_3) envisages space (or time) with a beginning and the metric reflects how far any measured interval is from this end point. Since the metrics are each adequate and intrinsic, none of them counts as conventional, *despite the different values they give to the size of intervals*. So the strategy of description and redescription, earlier the acid test for convention (Grünbaum (1968), p. 29) loses its significance. The reason for this, as Grünbaum explains it, is that these different metrics for discrete space tell different aspects of the spatial story.[2] M_1 tells the story of how many grains there are in an interval, M_2 tells the story of which grains make up the interval in question, and M_3 tells the story about where the interval in question is with respect to an end grain.

[1] I am neglecting distinctions between open and closed regions, for example.
[2] This metaphor of a metric's 'telling the spatial story' is Grünbaum's.

Each story is factual and none of the facts is inconsistent with the others. Since each story is both intrinsic and adequate, none of them is conventional. But no intrinsic story of continuous space is ever adequate. The plot is always too thin to be a metrical plot. So any story about its metric will be conventional.

These are the main ideas Grünbaum attributes to Riemann, developed a little in very simple ways. Clearly, the sketch fits quite neatly into the general conventionalist canvas we painted before. A primitive standard language tells us whether or not space is continuous or discrete, together with a number of other facts about how points 'make up' regions and how they are arranged in them. The standard language is as rich as topology, therefore, but no richer. This means that whether or not space is discrete is a simple brute fact which we can state in our primitive standard language. It is just because we can do so that we are able to go on later to view the metric of discrete space as intrinsic. But the standard language is not metrical: we cannot state it as a brute fact, expressed in primitive sentences, that such and such a class of intervals (or regions) are congruent. *We have to base it on something.*

In the next four sections I plunge into rather deeper waters in order to bring up the very detailed, rather technical development and justification of this view which Grünbaum offers in his article 'Space, Time and Falsifiability' (1970). All page references in the remainder of this chapter are to this article, unless otherwise stated.

3 Three kinds of metric function

We must first face a quite technical question: what is the mathematics of spatial measurement? Grünbaum's case depends heavily on how we answer this.

Mathematically, a metric for a space is some function which has entities of the space as its arguments and non-negative real numbers as its values. That is how we will regard it, at least. But this does not tell us just which kinds of function we should take as primary, which as most convenient, and how the kinds of function relate to one another. There are length metrics, distance metrics and measure metrics. Length metrics tell us about the *length of paths*, so the argument of any function which is a length metric is an arc, curve or path. A distance function tells us about the *distance between pairs of points* without direct reference to which path the distance is taken across. A measure metric measures *sets of points* and may be defined for a wide variety of sets of differing cardinal numbers and order types. Each of these functions can be used to define measurement for areas or intervals or regions (pp. 488–94). Which function should we choose?

Structurally, the simplest functions are distance functions $d(x, y)$, which take ordered pairs of points as arguments and non-negative real numbers as values. (Cf. Fine (1971), p. 452.) However, we do need to insist that what any distance function gives us is the length of the shortest path in the space of which the points are end points before we can properly take the distance function to describe a metric. Why is that? Well, take E_2 with Cartesian coordinates x, y given for it. Think of the subspace S_1 of those points such that $x^2 + y^2 = 1$. S_1 is the one-dimensional space of the unit circle (or one-dimensional sphere). Now take points P_1 and P_2 on the circle and consider the distance $d(p_1, p_2)$. Clearly, if we want d to give us a metric of E_2 the value of the function will not be the same as if we wish it to give the metric for S_1. In the second case we will want the distance across the shortest *arc of the unit circle* which joins p_1 and p_2.[1] In more complex cases, we will certainly want d to give us the metric which is intrinsic to the space *in the Gaussian sense* of 'intrinsic'. We want the metric *in the space we are concerned with* and must rule out a metric defined for a (possibly merely fabricated) containing space in which our space has some posture or other.[2] So, while distance metrics are defined on point pairs, they are relevant to the metric of a given space only because the distance is taken across a geodesical[3] arc joining the points.

Does this mean that length metrics are more fundamental than distance metrics? I believe it does, but we must be cautious about saying just how. In text books length metrics are defined in differential geometry and thus for differential manifolds with a rather exact continuity structure. This fits uneasily with the hypothesis that space might be discrete, and we could not use the integrals which define arc length in

---

[1] Van Fraassen (1969), p. 348. The construction put on the example should not be attributed to Van Fraassen who seems to regard length and distance as not clearly related.

[2] The Gaussian sense of intrinsic metric is not the same as Grünbaum's, of course. Gauss contrasts the intrinsic curvature, which is the geometry internal to a space, with the different curvature, or posture of the space with respect to some containing manifold. (See Chapter 3, §5, pp. 60–1.) Grünbaum is concerned, instead, with the contrast between a metric provided by a structure which is internal to intervals as against a metric provided by objects (e.g. rods) which are not parts of the space, though they are in it.

[3] The idea of a geodesic is not metrical, as will be explained in Chapter 9, §1.

the way we have them at present. Further, as Grünbaum points out, we should, perhaps, also realise that some length metrics are not distance metrics, nor do they provide a basis for them. In Riemannian spaces that have positive definite structure, length measures along geodesics give proper distance functions, but they do not do so in Riemann spaces generally. In Minkowski spacetime, for example, of signature $(+1-1-1-1)$ two light-connectible point-events lie on a null geodesic, so they would count as at no distance apart, despite being distinct points.[1] The mathematical structure of length metrics generally is more complex than we want, therefore. However, it remains true that the distance function, regarded as fixing the metric for a physical space, does take the distance as somehow defined across some geodesical arc or path. So length functions are basic despite these provisos.

Grünbaum places most weight on measure metrics. I will argue later that this is because they let him deal with the grains of discrete space in a certain way. But perhaps it is also because measure metrics do not depend on the complex structure of the continuum for their usefulness while, at the same time, they do take account of the structure of the sets to which they assign measures. Grünbaum applies them to discrete spaces, dense denumerable spaces and also to continuous spaces impartially. However, in other ways, they perhaps give us more than we want. Measure metrics do not apply only to intervals. The usual measure function on continuous space is defined so as to give a measure to the set of all rational points. In the closed real interval [0, 1] for example, the measure is zero.[2] So while the function is defined for intervals as sets of points it is also defined for sets of points which are not intervals. This appears not to matter much for discrete spaces, but it introduces complexities for dense or continuous spaces which we can profitably ignore. In what follows, we will use both distance and measure metrics, assuming in the first case, that the distance is taken over an appropriate arc and in the second case, that any set to be measured is either a singleton (a set with one point or grain as member) or an interval, unless I say not. I think that this is faithful to Grünbaum's practice and it is suggested, in large part at least, by what he has said.

[1] Grünbaum (1970), pp. 491–9 for this and related remarks. See also Chapter 10, §6.

[2] See Aleksandrov *et al.* (1964), Chapter 15, especially §4.

4 Adequate versus trivial metrics

We drew a simple distinction, earlier, between intrinsic and extrinsic metrics. Before looking more carefully at that, however, let me point out that the distance, length and measure functions allow some trivial results which, though they fulfil the axioms for metrics, are clearly not relevant to our worries. We need, then, to formulate some conditions for when these various functions are adequate rather than trivial. What they will be adequate to are the main aspects of our pre-theoretical intuitive understanding of what distance and length are. So they are conditions for the formal machinery to reflect our spatial concepts faithfully. The axioms for distance and measure functions (cf. pp. 488–9) are as follows:

A distance function may be defined, quite generally, as a real valued function d on ordered pairs of members of a non-empty set X, if and only if the function fulfils these axioms for every $a, b, c \in X$:

(i) $d(a, b) \geqslant 0$ and $d(a, a) = 0$.
(ii)   (Symmetry) $d(a, b) = d(b, a)$.
(iii)   (Triangle Inequality) $d(a, b) + d(b, c) \geqslant d(a, c)$. (See fig. 34.)
(iv)   If $a \neq b$, then $d(a, b) > 0$.

The real number $d(a, b)$ is called the 'distance' from a to b.

A measure function may be defined, quite generally, as a single valued function M on a set Y of subsets of X, Y forming a ring of sets, if and only if the function fulfils these axioms:

(*a*) The set function M is a *finitely additive measure* iff

(i)   for each member y_i of Y, $M(y_i)$ is either a non-negative real number or $+\infty$

(ii) $M(\Lambda) = 0$

(iii)   if the sets $\{y_i\}$ in Y are *disjoint* for different values

$$i = 1, 2, \ldots, n, \quad \text{then} \quad M\left(\bigcup_{i=1}^{n} y_i\right) = \sum_{i=1}^{n} M(y_i),$$

where it is to be understood that any sum is equal to $+\infty$ if one of the summands is $+\infty$.

(*b*) The set function M is a *countably additive measure*, iff M is a finitely additive measure *and*

(iv)   if the sets $\{y_i\}$ in Y are disjoint for different values

$$i = 1, 2, \ldots,$$

and if the union $\bigcup_{i=1}^{\infty} y_i$ of all sets $\{y_i\}$ belongs to Y, then

$$M\left(\bigcup_{i=1}^{\infty} y_i\right) = \sum_{i=1}^{\infty} M(y_i),$$

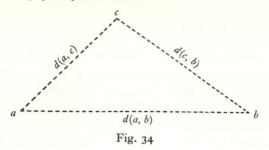

Fig. 34

where it is to be understood that if $\sum\limits_{i=1}^{k} M(y_i)$ is not bounded as a function

of k, then $\sum\limits_{i=1}^{\infty} M(y_i) = +\infty$, and also that any sum is equal to $+\infty$ if one of the summands is $+\infty$.

Clearly, these are very general definitions which apply to a far wider range of entities than those we want to count as spaces. There is no need for us to go beyond fairly intuitive ideas about what counts as a space. Our examples will always be pretty well behaved spaces.

Formally, we get a metric out of a distance or a measure function if it assigns the same value to every pair of distinct points or to every interval other than the singletons of dense or continuous space. Obviously, this alone will not give an adequate, non-trivial metric. For we found earlier, in our simpler account of the matter, that the proper inclusion of one region or interval by another ought to make the measures of the intervals different. This is an intrinsic fact about the intervals, after all. The first condition, therefore, is that any function giving a metric for a space will reflect the intrinsic facts of proper inclusion.[1]

Measure functions are not formally required to give any special status to singletons (unit sets with a point or grain as member). However distance functions always give the distance of a point or grain from itself as zero. No counterpart condition is formally laid on measure functions. Grünbaum states (pp. 509–10) that length functions must give any grain zero length. So he asks that if any set A properly includes a singleton, then $M(A)$, the measure of A, is not less than the measure of any singleton whatever (p. 501). It follows from these two conditions for the adequacy of the metric, that all singletons in dense space (including continuous space, of course) have measure zero. It also follows that the range of measure values open to singletons in discrete space is bounded above and below by positive finite measures. It does not follow that each

[1] Strictly, we need different values for closed intervals and their closed subintervals. We do not need a separate measure if we merely leave out the end points.

singleton of discrete space should have the same measure as every other.

Grünbaum considers whether a third condition is necessary for an adequate metric, though he rejects it. It is roughly the condition that the metric must differ for intervals which contain different numbers of grains (pp. 512–15).[1] If we accepted the condition, it would not mean that every pair of intervals with the same number of points or atoms must have the same measure, length or distance, *not even for discrete spaces*. Massey (p. 507) has provided an example of a measure function on a discrete space which is adequate under the three conditions just mentioned but which assigns a different measure to each singleton and to each interval; even so it gives greater measures to intervals with a greater number of atoms. The reasons why Grünbaum hesitates to impose this third condition are reasons of 'mere caution'. In a discrete space that has a boundary (an extreme or end element) it is possible to define a distance function which breaches this condition while being perfectly satisfactory, at least in Grünbaum's eyes, for other reasons. Finally, I point out that Grünbaum talks of differences among metrics being trivial if they differ just in scale; that is, by a constant factor. This is a useful idea that I put to work just as he does.

5 Intrinsic and extrinsic

Now we know, in a bit more detail, what it means for space to have an adequate metric. We still need a rather closer look at the contrast between the metric's being intrinsic to the space and its being imposed extrinsically. Grünbaum probes this at length (pp. 525–32). The core of the idea is enough for our purposes and it can be put fairly simply. A property is intrinsic to an interval if it belongs to the interval because of something inside the interval. A thing is inside the interval if the interval cannot exist without it. This is a rather specialised use of 'inside' – it puts an interval inside itself – but it is none the worse for that. We can extend this idea of intrinsic to relations generally in a pretty obvious way.

Grünbaum draws a useful distinction (pp. 527–8) between components

[1] It is obvious that this rough condition is not fulfilled for measures of point sets in continuous space. Both the set of rationals and any singleton have zero measure. Grünbaum's more careful condition is vacuously fulfilled by dense spaces.

of the metric: there is the equality component, the more-or-less component and the ratio-component. The equality component of a metric is an equivalence relation which partitions the set of intervals. This relation might be intrinsic to the intervals, but the other components not. Grünbaum gives examples of this in cases of discrete space, based on some suggestions of Massey (p. 539). Consider the measure functions M_1 and M_2.

(1) $M_1(I) = \bar{\bar{I}}$, where $\bar{\bar{I}}$ is the number of grains in the interval I.

(2) Let $S(a_i)$ be a 1–1 function from the set of grains, a_i, in discrete space to members of the series 1/2, 1/4, 1/8, 1/16... which sums to unity. Then

(a) $M_2(a_i) = 1 + S(a_i)$ for all grains;

(b) $M_2(I) = \sum_{i=1}^{n} M_2(a_i)$, for all $a_i \in I$ ($i = 1, 2, 3, ...$).

Here the equality components of M_1 and M_2 are both intrinsic, though they differ. M_1 is based on the equivalence relation, clearly intrinsic to intervals, of containing the same number of grains. By contrast M_2 is based on the equally intrinsic relation of identity: no distinct intervals or atoms have the same measure. But the more-or-less component (and the ratio-component) of M_1 is based on a partial ordering relation among the intervals, whereby $M_1(I_1) > M_1(I_2)$ if and only if I_1 has more grains in it than I_2. However in the second case there is no relation intrinsic to intervals I_1 and I_2 which holds if and only if

$$M_2(I_1) > M_2(I_2).$$

This gives us a very clear reason for preferring M_1 over M_2 as a metric for discrete spaces. It is as intrinsic as can be.

M_1 assigns the same measure[1] to every singleton in discrete space. This does not mean that we take the grains to be equal in size when we choose the metric. This must be stressed, since a number of Grünbaum's critics seem to have understood him to give equal size to the grains of discrete space in preferring M_1. This would then have to be a primitive property intrinsic to the grains. It is not clear that M_1 rests on any idea of this kind, however.[2] It takes the form we described simply so that it will meet the conditions for an adequate metric and so that all its components are based on intrinsic relations among intervals. It seems that M_1 is the only adequate and fully intrinsic metric for discrete space,

---

[1] I have made this measure 1, but this is just the simplest of many choices open to us. Since all these choices differ only by a scale factor, they are only trivially different.

[2] I will probe this whole question of the size of grains and the measure given to singletons when we take a more critical look at Grünbaum's ideas.

apart from trivially different metrics. There are strong reasons for preferring it, therefore.

None of this forbids us to define a metric on discrete space by some extrinsic means such as by carrying about a rod or by setting up of some kind of general apparatus of measurement, making use of caesium radiation for example. There is no reason to think that metrics with a basis like this will not be formally in order, nor is there any reason to suppose that they will be trivial. They will simply be extrinsic.

Grünbaum does not claim that dense or continuous spaces never have intrinsic metrics. They certainly do all have them but these intrinsic metrics are trivial in every case. He provides two examples. The first assigns to every singleton the measure zero and to every (non-degenerate) interval the measure $+\infty$. This measure clearly reflects the intrinsic facts about the number of points in the different kinds of sets. But it fails to reflect other intrinsic facts about the inclusion of some intervals by others, so it collapses into triviality. The second example is even less adequate: it gives the measure zero to every set, singletons and intervals alike. Grünbaum takes this measure to be intrinsic since we could base it on the property of each set that it is identical with itself (pp. 546–7). Two sets are equal in measure (o) just if each is the same as itself. Every set measures the same as every other.

Grünbaum does not take for granted that the only properties of a manifold which we could use as a basis for an adequate metric are exhausted by those that tell how the interval is made up out of points or grains. He shows that in the case of the real numbers we can define various metrics which are both adequate and intrinsic. Even though real numbers go to make up intervals in just the same way as points go to make up intervals of continuous space, the numbers differ from one another intrinsically just in being the different numbers that they are. To illustrate, look at these two measure functions (p. 533):

$$M_i(a, b) = |a-b|$$
$$M_j(a, b) = (a^2 - b^2) = (a+b)\,|a-b|.$$

In each case the measure is based on these equivalence relations, clearly intrinsic to intervals:[1]

$R^i(A, B) \equiv$ the absolute difference between the end points of A = the absolute difference between the end points of B.

$R^j(A, B) \equiv$ the absolute difference between the squares of the end

---

[1] Obvious adjustments would be needed to deal with open and half open intervals.

points of A = the absolute difference between the squares of the end points of B.

Both these relations are intrinsic to the intervals A and B, because each interval contains its end points and it is intrinsic to these points that they are just the real numbers that make their difference take a definite value. Spatial points are indistinguishable so they have nothing to base a relation like these on. We could give them coordinate real numbers, of course, but they would be extrinsic.

All of this raises the question which metric to choose, given that spaces offer us a choice among several. Clearly, we can rule out trivial metrics without more fuss, leaving only extrinsic metrics for continuous space. But where there is an adequate intrinsic story to tell 'the object of a metric... *is to render intrinsic facts or tell* the *intrinsic* story *in so far as possible*' (p. 576). So for discrete space, intrinsic metrics win the field over extrinsic ones since they have a story worth telling. There may be different stories to tell, so there may be different metrics to tell them. Though these stories are certainly factual ones, they are not competing, inconsistent stories, just different ones. But, as we said, one of these, the M_1 story, is the most detailed, so we prefer it.

6 Different metrics and conventions

What we have rather clearly before us now is the question whether the alternatives show that the metrics are conventional rather than factual and how they show it. In the latest statement of his view on the question of convention, Grünbaum makes it clear that he ties the question of what is conventional to the question of what is intrinsic. We ought to notice three aspects of his ideas. First, we can give different adequate metrics on one and the same discrete space without any of them being conventional. Therefore, second, we cannot show that a metric for continuous space is conventional just by showing that different adequate metrics can be given for it. Third, suppose that only continuous space permitted different metrics. This would still not get at the core of what frees discrete space from metric conventions. The core of the matter is that discrete space is so constructed that every interval has an intrinsic basis for its equivalence and ordering relations, whereas continuous space is not constructed like that.

We have already looked at two of these metrics, in examining the measure functions M_1 and M_2. But suppose we have to deal with a discrete space (or, more suggestively, a discrete time) which has a beginning – a boundary or first grain. Then we attach coordinates to successive grains in an intrinsic way defined by the betweenness relation on the space (p. 486). We can give a distance function on the space which is intrinsic *to the space*, though not to the intervals generally since it depends on reference to the first grain. This makes the distance a function of the number of points in the closed interval, but also of how close I is to the first grain. Now, this is not a conventional metric, according to Grünbaum, since it is intrinsic to the space (time). Though it differs from M_1 and M_2 for the same space, it does so by telling a *different but compatible* intrinsic story. Grünbaum puts the matter as follows: 'I know of no *overriding* reason to consider sensitivity to cardinality more informative spatially in this context than the stated sensitivity to ordinal proximity to the end-point' (p. 513: see also pp. 544–6). Finally, let us recall that it is quite possible for us to give an extrinsic metric on a discrete space by using a transported rod. In fact, we may be forced to do so if the discrete atoms are too small to count effectively.

Just why are differences among these metrics not reckoned to prove them conventional? We must remember that it is said that the first duty of a metric is to 'tell the intrinsic story' of the spatial facts. M_1 and M_2 tell factual stories that differ *without competing*. One tells us how many grains in an interval, the other tells us which interval it is. There is nothing conventional about this. Again, the metric M_2 tells us nothing by way of an intrinsic story in its order and ratio components, though its intrinsic equality component does tell us which intervals are which. Perhaps, then, we should see M_2 as tinged with convention to the extent that, as an alternative metric, it misses a bit of the spatial story. But the extrinsic rod-based metric, which we are so familiar with, tells no intrinsic story, except for the trifling one of which interval includes which. Since this story is available to us in other ways, the rod-based metric is conventional (though adequate). The same is true of any metric in continuous space.

There is the possibility that some physical theory might give a unique metric to continuous space or spacetime as part of its description of the world. Grünbaum agrees that General Relativity is just this kind of theory. Does this mean that the metric is factual and not conventional after all, if the world is as this theory says? Grünbaum argues that it is conventional because it is not intrinsic even though it is normative (prescribed by the theory). As he points out himself ((1970), §3(i) and (ii)) this seems to cut any very useful link between convention in a metric and metrical amorphousness on the one hand and the possibility of different metrics for a space on the other. The earlier strategy of description and redescription has almost completely gone.

Seeing the metric as part of a global theory of physics calls for more reflection, however. Riemann said that we need some means of 'carrying a magnitude forward'. The obvious idea is to carry about a rod which, we stipulate, delivers the same length wherever it goes, at all times. The stipulation is a convention. The convention is only an ingredient in the metric, since it is a matter of topological fact whether or not the rod overlaps a certain interval at a certain time. But a glance at what actually goes on in measurement might make us wonder how realistic this is. Certainly it is far more complex in practice even if not in principle. Very elaborate and expensive precautions are taken with standards like the metre bar in Paris to prevent them from 'varying in length'. Another worry is that it is not clear that we have to carry a rod, or anything else, from one interval to another to get an accurate comparison of lengths. We can set up similar bits of apparatus in the two places to measure and compare lengths using radiation. This does not involve transport even in principle, though it certainly involves theory. However, in all these cases, the procedure for measuring lengths is extrinsic. This might seem to be enough[1] to justify Grünbaum's regarding theoretical metrics as conventional, provided that we grant the rest of this position. But it is important to see that the metric will be *caught up in the content of a physical theory*. We make the metre bar of platinum-iridium and take special precautions about its temperature because our physics tells us that it will change its length if we don't.

This is a bit too quick, however. We will have to recognise, too, that physical theory is tainted by convention. We noticed earlier that Reichenbach saw the conventions as dealing with universal forces – those that are supposed to affect all substances in exactly the same way. This makes the dynamics of physical theory a convention. But this is not Grünbaum's way with the problem. In (1964), Chapter 2A Grünbaum looks at two metrics for time.[2] He claims that they are equally adequate to describe the objective physical facts and to explain them. The first time-metric is the one used in classical Newtonian physics, the second is determined by the rotation of the earth, the sidereal day. The two times

---

[1] I do not think it really is enough, and I say why in §1 of Chapter 9.
[2] Grünbaum has several times discussed other cases of alternative time metrics that have actually been proposed in physics ((1964), pp. 77–80, pp. 22–3; (1968), pp. 70–80). These will be discussed in the context of relativity theory. (See Chapter 10.)

are not related to one another in any very simple way; in particular, they are not linearly related. Things that move uniformly according to one metric will not do so according to the other. This might seem to mean that the dynamics of the two theories would differ and that we would have to write this difference off as conventional, in Reichenbach's manner. For example, the Newtonian theory can easily explain the slowing of the earth's rotation in familiar ways that link it with the gravitational action of nearby masses, for example. There will be no such slowing down for the other time-metric, of course, but some bodies will be described as accelerating which have no Newtonian acceleration. How shall we explain these? Grünbaum suggests that we are not obliged to find a dynamical basis for acceleration in every possible theory. In the new theory, a certain class of accelerations are seen as force-free motions and they correspond with Newtonian uniform motion. There is no need to bring in, as conventions, any new bits of dynamical machinery. We can simply reformulate the laws so as to relate force and motion a little differently. Newtonian formulations are simpler, of course, but Grünbaum sees this just as descriptive simplicity, not as a greater adequacy to report facts or give explanations.

7 An outline of criticisms

Despite the rather elaborate, ingenious and technical array of ideas laid out in the last four sections, I think that Grünbaum's explanation breaks down for rather simple reasons. One way of showing this is to make my point in two stages. First, I insist that intervals of discrete space just divide up into grains, which are also intervals of the space. That is, any interval of discrete space is extended because each grain is extended. How far an interval in discrete space is extended is simply a matter of how far its component grains are extended. An interval's dividing up into grains is a naive business, not a tortuous set theoretical one. It means that a discrete interval has an extension and a size intrinsically and factually if and only if *each grain has an extension and a size intrinsically and factually*. Since the metric is granted as intrinsic to discrete intervals it must be intrinsic to the grains.

The second stage of making my point is to insist that continuous space is like discrete space in the crucial respect that any interval of it divides into other intervals of it. As before, a continuous interval has an extension and a size intrinsically and factually if and only if each subinterval has an extension intrinsically and factually. Now if the un-structured grains of discrete space have an extension and size, as a matter of intrinsic fact, there is no reason whatever why intervals of continuous space should not have extension and size as a matter of intrinsic fact. That they are 'made up' out of infinite sets of points in some way or other does not detract in any way from this. For points do not add anything to the extension or size of intervals. So there can be a metric, free of conventions, for continuous space.

There is nothing particularly novel about this kind of objection to Grünbaum's view. Both Earman (forthcoming) and Swinburne (1970) have already pointed out that it must be no less arbitrary to assign sizes to the grains of discrete space than to assign sizes to the intervals of continuous space. As I understand him, Grünbaum's immediate reply to this objection is to damn as naive the suggestion that grains have a size. What has measure, according to measure theory, are sets, so it is always a set of grains (perhaps just a singleton set, of one member) that has a measure. But for this reply to work, we would need to admit some parallels between the way an interval of discrete space is made up of grains and the way an interval of continuous space is made up of points. However, the parallels fail, as I will argue in detail.

I think that Grünbaum nevertheless takes his reply to succeed because he takes it that conventionalism really does not need to be justified, but only explained. None of his books offers a real justification of conventionalism. He takes its essential doctrines for granted. Thus, in the present case, it is assumed right from the start that there is some distinction between intrinsic or primitive properties and extrinsic or derived ones. The topology of space (its being discrete, for example) is primitive. It is not based on some other property of space. It is asserted to be discrete as a matter of plain brute fact, and that's that. We can just say what the topology is, for the sake of our argument, and press on. The metric is simply assumed to be a different kettle of fish. It is taken to be neither primitive nor intrinsic. We may not just say what *it* is, for the sake of our argument, since it is assumed at the outset that *we have to base it on something else*.

Grünbaum needs this contrast since a metric can't be intrinsic to discrete space unless its topology is. Nor can continuous space be intrinsically metrically amorphous unless *its* topology is intrinsic. Without this contrast the game never gets started. But the contrast comes from nowhere. It is not epistemological since the argument is never directed toward our conceptual limitations but always at limits upon the structures things (intervals) can assume. But why can't a metric be intrinsic and primitive, too? Of course, a metric is a further primitive determination of a space beyond its topology. But that does not rule it out as primitive and intrinsic.[1]

However, the issue is not just that Grünbaum takes for granted something which I believe we ought not to grant. It is that he brings to the problem convictions about what it is permissible for a metric to contain that are deeply contrary to our intuitions about size and measurement. Despite what he says about discrete space, I think Grünbaum really believes that any metric is in some sense a convention.[2] That is, there is no primitive idea of spatial quantity which a metric has to be adequate to capture. What is required of a metric is pretty meagre. So there seems to him no harm in feeding into the metrical story a number of things which are really repugnant to our intuitions about size and quantity in space. Thus we get quite different 'intrinsic' metrics for the same space. In one spatial story there are elements of plot about *how many* grains are contained in an interval (which is emphatically not meant to reflect a size which the grains have); in another story the plot tells us *which grains* are in the intervals; yet another tells us *where* the interval is relative to an end point of the space. But, I submit, any metrical story has to be simply quantitative. It is to tell us about size. None of Grünbaum's metrics really reflects size and only the first even looks as if it does. Unless someone shows that there can be no simply quantitative facts about space, we should reject these metrics. Grünbaum has not shown it.

The remaining sections develop key elements of this criticism of Grünbaum in detail: §8 looks into the division of intervals in discrete and continuous space, noting parallels and contrasts; §9 looks into Grünbaum's ideas about whether grains of discrete space have a size; §10 argues for the importance of the parallel in the division of both discrete and continuous spaces into intervals; §11 argues for the unimportance of the contrast between the way discrete and continuous spaces are made up of grains and points, respectively.

[1] This is discussed at length in Chapter 9. It is perhaps worth adding here that what is primitive is intrinsic. A property is extrinsic to a thing if and only if whether or not it applies is based on something external to the thing. A property is primitive to a thing only if whether or not it applies is not based on anything but the thing. *A fortiori* its applying is not based on anything external.

[2] I was convinced of this after a very valuable discussion of the whole argument with Bas van Fraassen. I am not sure that this was what van Fraassen really wished me to conclude and he is certainly innocent of sharing my opinions about the issue.

8 Dividing discrete and continuous spaces

Let me begin the detailed critical tasks by offering my own account of what the grains of discrete space contribute to its metric. It is a quite different account from Grünbaum's, very much simpler and, I think, obviously correct. It might not seem to be so very different from his, at first, but I believe that impression will not survive a close look at what he says and careful thought about what he needs to say.

Take a spatial interval and consider ways of dividing it. It may be the case that however we try to divide it we arrive at grains or atoms which we can divide no further. But it may be the case that we can divide it into proper parts in such a way that we would always be left with intervals which we can divide further. That is, we can find processes of division that do not end and which are not trivial processes. In the first case, the interval is a discrete space, in the second it is dense and, perhaps, continuous.

Let us think about the first case first. We reach indivisible grains in a finite number of steps. What you reach, if you finish a process of dividing an extended interval, are extended grains which will not divide further. No matter what is said about whether the metric is intrinsic, we must surely say that any spatial interval is intrinsically extended and that its extension is a primitive property of it. Spatial grains are intrinsically and primitively extended too. In fact, the grains of discrete space are themselves intervals, since they are simple parts of other intervals, even though they are not divisible into further parts. *Discrete space and continuous space are alike in this respect: they divide into intervals which are their parts.* They differ only in that what continuous or dense spaces divide into are always further divisible. But in both cases, quite obviously and simply, they divide into interval-parts, each of which is extended in a primitive, intrinsic way. To look at the same idea from a different direction, an interval of discrete space is a sum or whole of its atomic, granular parts. Intervals of discrete and continuous space are sums of interval-parts in just the same ways. For discrete space, there are ultimate parts (grains); for continuous space, each part has further parts. In neither space can we speak of division into members, for the intervals of discrete space are not sets but wholes of their parts (grains) whereas the parts of intervals of dense spaces are subsets, not members, of the intervals that contain them. As for points, on the other hand, an interval of dense space is a set of points, not a whole or sum of them.

What lies behind this last statement? It is a fact that continuous space is infinitely divisible. This does not mean that points lie at the end of an infinite process of division which, unhappily, can never actually be completed. Infinite processes do not end, nor do points (or anything else) lie at the end of them, therefore. If space is infinitely divisible then it (properly) divides only into intervals without our ever reaching an indivisible one. The only parts of continuous or merely dense space are its (proper) intervals; it is a sum only of these. Points are not parts of space nor is space a sum of points. Points are members of intervals, not parts of them. At the risk of tedium, I will repeat that *intervals never (properly) divide into points but only into progressively smaller intervals.* The set theoretical treatment of points and intervals is not a perverse bit of mathematical pedantry nor due to the obscurantism of those who insist on toying with the intellectual machinery of their own discipline. It is important to see that it is false that points are parts of space, or that (proper) intervals are wholes or sums of points.[1] Certainly points are in space; they are boundaries or positions in it. We are sometimes encouraged to think of points, curves and so on as idealisations of familiar objects; pencil dots or strings. But as W. K. Clifford (1955) pointed out, points, curves and surfaces are boundaries and not abstract. The boundary between the table top and the air is a surface: it is where the wood ends or the air begins. If a surface is dipped in water, so that it is half-wet and half-dry, some curve is the boundary between the areas: it is where the dry ends or where the wet begins. A point is a boundary on a curve where some feature changes, colour, say: it is where the red ends or where the green begins. But points are in spaces as members are in sets, not as parts in wholes. Intervals are made of smaller (proper) intervals in rather simple ways. We might suppose that they must be sums of points only in some very complex and tortuous way. The truth is that they are not 'sums of' points in any way at all.

This is a bit incautious. To make the theory of the real line very general, it is useful to include among the intervals a point set which contains only a single point. An interval is the ordered set of real points between any pair of points on the line. The degenerate case of the closed interval is given when $a = b$, so that the interval is $[a, a]$ or $\{a\}$, the singleton, an 'ordered' set containing one point. Now we can certainly reach degenerate point sets by degenerate division. From the closed

---

[1] Perhaps facts about infinite divisibility do not force us to a set theoretic account of intervals, but the account certainly lets us make the distinctions we need with clarity and ease.

interval, [0, 1], subtract the half-open interval [0, 1) or the two half-open intervals [0, $\frac{1}{2}$), ($\frac{1}{2}$, 1]. The first 'division' gives us the singleton {1}, the second the singleton {$\frac{1}{2}$}. Clearly, this is something quite different from the intuitive idea of division which leads us to say that you cannot divide a line into points however far you go dividing it. To say that the intervals of continuous or merely dense space are infinitely divisible and that points do not lie at the 'end' of processes of division does not mean that we have no way of reaching an indivisible degenerate interval immediately. But this is not dividing the interval into its parts. Of course the proper interval is the union of the uncountable set of singletons which exhausts all its degenerate subintervals. But singletons do not contribute to the extension of the interval; they do not 'add up to' the interval in any sense. The measures of the sets [0, 1] and [0, 1) are the same; the measures of [0, $\frac{1}{2}$) and ($\frac{1}{2}$, 1] add to the measure of [01]; the measures of {1} and {$\frac{1}{2}$} are zero. The extension of intervals in dense or continuous space does not derive from some property of points or singletons; neither from their number, their order nor, of course, from their extension, for they have none. So, in a perfectly non-arbitrary way, the proper interval is not the sum of the singletons (let alone of the points themselves) and they are not contributory parts of it.[1]

What this story of spaces and their parts depends on is the same assumption I made before, which is that spatial extension is a simple, primitive idea. Spatial things and space just are extended, intrinsically and primitively; things just are separated across spatial arcs, paths and so on.[2] If a metric can be intrinsic to a space, then certainly extension must be. If discrete space is intrinsically extended then its simplest parts must be intrinsically and primitively extended, since the extension of an interval is just the sum of the extensions of its granular parts.

In sum, I want to emphasise that discrete and continuous spaces are alike in dividing into intervals, whether the intervals are grains or not. Any interval of either space is the sum of the (non-overlapping) intervals it contains. I also want to emphasise that there is no significant parallel between the way discrete intervals contain grains and the way continuous intervals contain points.

[1] Grünbaum (1964), p. 169.
[2] See Chapter 2, §6, on pathological adding.

9 Discrete intervals and sets of grains

There is rather a lot of evidence in 'Space, Time and Falsifiability' that Grünbaum does not, and I think, cannot, share this simple view of extension in discrete space. But the evidence is perhaps equivocal. I am not sure that Grünbaum directly intended to reject the simple view, yet I believe that his defence against certain objections is inconsistent with holding it. He approaches the metric of discrete space through the apparatus of measure theory. This way of dealing with the problem is possible according to the simple view I am urging, though it is much more laboured than need be. In measure theory, one always measures sets, usually sets of points, of course. But, if I understand him, Grünbaum's ideas depend on this set theoretical approach since only it lets him deal with the problem of grains in space as if it were the problem of measuring the extension of singletons – unit sets – with a grain as member. Without this elaborate, and I think confused, treatment Grünbaum would be open to the objection that assigning a length to the grains of discrete space is just as arbitrary or conventional as it would be to assign lengths to intervals of continuous space. That will be the theme of the rest of this section.

Let us probe a bit deeper into the objection just touched upon. Several writers have voiced doubts about the idea that the metric *is* intrinsic to discrete space. It seems to involve either an assumption or a convention that the grains are all the same size. Unless we either assume this or stipulate it, the metric is indeterminate whatever the number of grains in an interval might be. It can be neither more nor less reasonable to assume that grains are intrinsically the same size in discrete space than to assume certain intervals are intrinsically equal in continuous space. Grains cannot be moved and matched one against another any more than continuous intervals can. Of course, the simplest, most convenient thing is to take all grains as equal. There is no question of our doing anything else. But simplicity and convenience also work out a unique metric for continuous space. This is the metric we actually use. So, the objection goes on, Grünbaum's introduction of a new concept of convention in metrics has failed in its first task: it has not made out any distinction between intrinsic and extrinsic metrics. Arguments along these lines can be found in d'Abro ((1950), pp. 40, 40n) where Grünbaum is not the target, of course, and in Swinburne ((1970), pp. 310–11) and Earman (forthcoming), where he is.

An obvious way to avoid this objection is to see if you can argue that a discrete metric assumes no hypothesis and stipulates no convention about the size of grains. Grains have no size. Though Grünbaum does not say this in so many brisk words, it does appear that he thinks it is the case. When he deals with the length of singletons in discrete space ((1970), pp. 509–10) where he is not directly worried about the objection, he takes it that the '*length* of any singleton *is zero*' guided, so far as I can see, by the fact that the distance of any grain from itself is zero. He says that, in discrete space, the measure and the length of intervals are not as simply related as they are in continuous space. He concludes that 'the non-trivial measures of intervals cannot be equal to their lengths' and that 'length and measure of intervals are related in other ways, if at all' ((1970), *ibid.*). But if an interval has length and the interval divides into grains, as it does in discrete space, then the grains cannot have zero length, for they must simply add up to the length of the interval. It is beyond me to grasp what concept of length would permit any conclusion other than this simple one. Of course, length is not quite the same thing as measure, but this is a first indication that Grünbaum does not see the role of grains in discrete space in the simple and direct way I recommended in the last section.

When Grünbaum comes to tackle this objection head on, his tactics are not easy to understand. His direct reply (pp. 574–5) is that there are various metrics for discrete space which he thinks are both adequate, and intrinsic. Apart from his own front runner in measure theory, M_1, where we just count the grains, there is Massey's 'progressive' measure function, M_2 and a metric M_3 defined on discrete space with an end point. But what can the aim of this reply be? Not that we are free to make other assumptions about the size of the grains: that would be to admit the objection. The reply appears to be that these different metrics are founded on different, but not competing, intrinsic facts about *sets of grains*. The central idea seems to be that grains are the wrong kind of thing to have size, and that critics have naively supposed that a metric could ever give size to a grain. Grünbaum bases these various metrics on different equivalence relations among intervals. But these equivalences have nothing to do with equivalences in respect of the *size of grains*. They are equivalences founded on set theoretic features of intervals, such as having the same cardinal number of member grains, or having identical members, containing the kth grain, and so on. The size of a grain is never allowed to count. So the measure given to a singleton (a set) is not supposed to be applied to, nor is it derived from, the size of the grain which is its only member. Nor is measuring an interval by the cardinal number of its grains a way of adding the extensions which the grains have primitively.

This is an extremely sophisticated idea; perhaps my interpretation seems a farfetched one. I will quote Grünbaum:

'Thus M_1 and M_2 do not *disagree* in regard to any intrinsic properties of intervals. And hence it would be altogether misleading and wrong-headed to characterise the differences in the numerical assignments which these measure functions make to singletons, for example, as reflecting different "assumptions" about the "sizes" of the singletons in the sense that at least one of them might be mistaken' (p. 575).

As I pointed out before, this cannot mean that different assumptions do not compete because assignments of size are conventional. Nor can it mean that the metrics are trivially different, since they differ in more than a scale factor. It has to mean that assignments of size are not made to grains directly or primitively. It means that different metrics single out different (but compatible) features of singletons as sets. If one of the features were allowed to be the size, the quantity of extension of the grain, then it would surely take precedence over the rest, metric being precisely measure, length or distance. It is measure of extension. Grünbaum's assignment of measure to singletons, then, does not reflect the length of the grain. The length of the grain was said to be zero (pp. 509–10) and only the set can be given a non-zero measure. I take it that these are the reasons why 'assumption' and 'size' appear in scare quotes in the passage just repeated. The upshot of this section is as follows: Grünbaum directly denies length to grains. Less explicitly, he seems to argue that, though there will be things with smallest measure in discrete space, it is a confusion to think that these are grains. They are sets of grains: singletons. Only this last argument would seem to provide a way of meeting the objection we began with.

10 Grünbaum and the simple objection

This reply to d'Abro and the rest is too subtle to be true, however. The plain brute fact is just that grains in discrete space are indivisible intervals, they are parts, not members, of larger intervals. They are themselves extended if the space is. Neither d'Abro, Swinburne nor Earman seems to bring this point out clearly, however.[1] The size of an

---

[1] Perhaps this is because they do not see how elaborate Grünbaum's point really is.

interval of discrete space does not come from some feature of intervals a sets of grains. It comes straight from the size of the grains by adding them. The interval can have no size unless the grains themselves have it. Since the grain must be extended if the interval is, one being part of the other, it *adds its length* to others in making intervals up. Metrics of discrete space that differ in more than trivial ways do disagree about the size of the grains of space. The only thing that can properly be understood by 'the measure of a singleton' is the size of its member grain: how much this part of the interval adds to the whole. Nothing else about singletons in discrete space is metrically relevant. The intrinsic extension of grains cannot be spirited away by ingenious stories about features intrinsic to sets.

So far as I can discover, Grünbaum always expresses himself at least consistently with this too elaborate view. But he never flatly rejects the simple view anywhere in his work. I take two more examples. Let us focus on the way he lays down the first condition for the adequacy of a spatial metric. Roughly, the condition is that we want the metric to reflect the intrinsic fact that some intervals include others. Grünbaum's statement of the matter is a bit more careful and circumspect: 'Hence to say that (a, c) *extends over* (a, b) *by a half-open interval* is to say that all of the members of (a, b) also belong to (a, c), but that (a, c) includes a (half-open) interval whose elements do not belong to (a, b)' (p. 496). The idea of one interval's extending over another is carefully explained by means of the set theoretical notion of subset, not the simple idea of one interval's being inside another as a part in a whole. This is certainly not a clear rejection of the simple view, however. When Grünbaum comes to state the condition it appears (p. 497) as:

$$(x)(y) \{(x \subset y).(x \neq y) \supset M(x) \neq M(y)\}$$

The measures are simply required to be unequal. The containing interval does not get a greater measure because it is just plain bigger and has the other interval inside it. It is rather because the axioms for finitely additive measures on rings of sets will require it, once we grant the conditions of adequacy. So Grünbaum moves towards M_1, which is the obvious and 'natural' metric for discrete space, by a series of steps. No step takes a spatial grain to have an extension or size which is part of the extension or size of an interval. Every step treats the interval as a set of atoms, earning its measure out of set theoretical features which it is argued to have. Grünbaum bases these various metrics on different equivalence relations among intervals. But these equivalences never hold in virtue of the *size of grains*. So, in one place where he might well have stated the simple view, if he holds it, Grünbaum has repeated the elaborate view instead.

The second adequacy condition is probably more interesting. It is a condition about singletons of any space, either dense or discrete (p. 501). It guarantees that singletons of discrete space always get a smaller measure than any other interval.[1] So it is highly relevant to the objection we were just looking into and highly debatable for that very reason. First, let us look at a discussion in which Grünbaum leads up to a statement of this condition. He takes it for granted that we face the same problems with grain singletons in discrete space as with point singletons in continuous space. He motivates this second condition of adequacy by arguing that a measure function on the manifold of natural numbers might tell one of two intrinsic stories. The first story requires the same measure for every singleton, no matter which number is its member. This tells the tale that each singleton is like every other in having just one member. The second story gives the measure n to the singleton $\{n\}$. This tells us which element it is that the singleton contains (p. 499). But for spatial manifolds, not only is every singleton like every other because it is a unit set, each member is like every other in being indivisible. The second adequacy condition for a spatial metric requires that this fact about singletons (indivisibility) be reflected in the measure function. In full: For any space 'if a set A properly includes a singleton, then the measure $M(A)$ of A cannot be less than the measure assigned to *any singleton whatever*' (p. 501). So the problems of measures for singletons look the same for discrete and for dense or continuous spaces.

This says nothing about the size of grains. It might seem to say that the size of grains in discrete space is never greater than the size of divisible intervals. But it does not say this. It says something about sets and members, not wholes and parts. A singleton gets smallest measure. That is because its only member is indivisible. *It is not because a grain is small, which is an utterly different idea.* I do not think I am riding an unlucky terminology unfairly hard. I am making Grünbaum's point not marring it. He does say that singletons are *unextended – always – even if they get a non-zero measure in discrete space.* Finite additivity demands a non-zero measure for singletons or else the metric is trivialised. This does not give grains a size. Their length is alleged to be zero. How else can we take the following passage, which is about spaces quite generally? 'Thus, while singletons are the only sets that are indivisible in the

[1] But, of course, they still get a length of zero.

specified sense, they need *not* be the only sets...that are *metrically unextended* in the sense of being assigned the *lowest* among the non-negative measure numbers assigned to non-empty sets of that space' (pp. 500–1). What is intriguing here is the sense that is given to 'metrically extended'. The sense seems clearly intended to entail that the singletons of discrete space *are not extended*. This has to mean, as I see it, that an interval in discrete space is extended *only if it has non-empty proper subsets*. If this interprets the text correctly, then the set theoretical treatment of intervals in discrete space is very different from the simple and obvious part–whole treatment. It then seems certain that Grünbaum tries to ward off the objection that he makes assumptions about the size of grains by denying sense to the idea that they have any. But the set theoretical treatment is not a sound defence against the objection, since it is confused. The plain facts are that the grains of discrete space are themselves extended: that their being extended is brute, primitive, and intrinsic, and that extended intervals divide up only into extended intervals. The resulting intervals can be sets of points or they can be grains: they can never be sets of grains. The measures and lengths of intervals in discrete space do not spring from the measures of unit sets, measures which are given to them just to tell the story that their members are indivisible and to tell it in a metrically adequate way.

11 What metric stories are about

Once we concede that grains are parts, and are not members, of intervals, Grünbaum's second condition for the adequacy of a measure metric looks all wrong for discrete space. Of course, the whole measure theory approach to discrete space is unnecessarily elaborate, and even conceptually off-beam since an interval of discrete space is precisely *not* a set of its grains.[1] But if grains are simple parts of intervals, the first thing to settle is how long each grain is. Until we settle it there is no reason at all to be concerned, either with the fact that grains cannot be divided or with how many of them there are in some interval.[2] If grains are what

[1] We could say this: an interval has a measure which it derives from the measure of the set of grains which it divides into. There is every reason of simplicity why we shouldn't say it, but it would be conceptually congruous.

[2] The remarks about the numbers of grains in this paragraph go beyond what is strictly relevant to Grünbaum's second adequacy condition, though certainly not to what he says generally about the number of grains in intervals. I include them together since the one argument deals neatly with both these mistaken emphases.

intervals divide into, then grains must have a length and at least some have a length greater than zero.[1] But I can think of no reason at all, either in or out of Grünbaum's pages, why we should not suppose that grains differ widely in size. Nor can I think of any reason why we should not suppose, further, that some single grains are longer than some intervals which contain several grains. There can be a simple relation between the size of an interval and the number of grains in it only if all grains are equal. Even then, it is the equality of grains that is the primary metrical fact. So Grünbaum's second condition for the adequacy of a metric could seem right only if we confuse intervals of discrete space with sets of grains. To stress that grains are not divisible misses the metrical point. To emphasise indivisibility or cardinal number is surely quite misguided here.

In a quite different way, it would seem equally wrong to worry ourselves about the cardinal number of points in intervals of dense or continuous space. Since intervals do not (properly) divide into points, not even by 'infinite division', we have no reason for thinking that points contribute in any way to the length or the measure of intervals. Of course, the question of having a measure, and thus of contributing one, can only arise with point sets, in dense space. That is, it arises primarily with intervals, whether degenerate or not. Still, it seems quite wrong to think of the cardinal number of a point set as having a central metrical significance. True, the axiom that the measure of sets be countably additive ensures the measure zero for any finite or countable point set of continuous space. Non-zero point sets are always uncountable in continuous space, but it is not this simple difference of cardinal number that works the trick of making non-zero point sets. There are uncountable sets with measure zero, for example, the Cantor 'third' set (Courant & Robbins (1941), p. 249). In fact, measure theory seems best understood as the theory how to take a primitively understood non-zero measure for simple intervals and extend it to point sets derived from these intervals by operations such as union, intersection, complement, etc. So the task is not to arrive at an understanding how to measure intervals from facts about points and unit point sets, but rather the reverse.[2] I believe it to be a mistake to suppose that intervals achieve non-zero measure in some way derived from either the cardinal number

---

[1] I think the idea of a zero-length grain in discrete space is senseless, in fact, but I see no need to make my argument depend on that.

[2] I realise that measure theory is more general than these remarks suggest. I mean only that this is where the theory begins.

or the order-type of point sets. Therefore, it seems a misunderstanding to suppose that if a metric is to 'tell the spatial story', then it will tell us a lot about cardinal number. That turns out to be quite peripheral to the spatial plot.

If a metric is to tell a story about what is intrinsic to space then there is only one story it can tell, the story about extension, size, length or distance. If we build other elements of plot into it, then surely it is no longer a metric. As I said before, a metric in discrete space has to tell us *how big grains are*, not that they are indivisible. That a thing is indivisible is simply irrelevant to its size: it may be a point or an extended grain. That a thing is extended tells us nothing about whether we can divide it or not: they are just different ideas. But further, Grünbaum apparently sees no reason why a metric should not tell us about where a grain or interval is when it is being measured. The simple fact is, however, that these bits of information are not about extension, size, length or distance: they have nothing to do with how big the grains or intervals are. Length and distance are not concerned with the identity and position of things. It is really very obvious that these concepts are quite distinct.

Let me close the chapter by generalising the moral to any kind of metrical conventionalism. The conventionalist theories begin with the idea that length cannot be a brute, primitive fact about an interval if space is as we usually assume it is – continuous. But this raises a problem about what on earth the metric is doing in our theory of the world. If the metric is an imposed convention, and factually idle, it is quite inscrutable why we have imposed it, as I argued in Chapter 6. We can easily see that some conventions are simpler than others, but it is far from obvious that any convention is simpler than none. In fact, it seemed before that a convention-free language is inevitably simpler than a convention-bound one. It is doubtful how it can make any kind of sense to have a metric for space if the metric is a convention. For a conventionalist, one way out of this wood might seem to lie in arguing that a metric can give us a factual report on its space *in some cases*. So we get the initially appealing idea that a metric gives a factual description of discrete space.[1] But, actually, any simple way of applying the principal metrical ideas – size, length and distance – to discrete space immediately suggests a way of applying them to continuous space. So Grünbaum avoids the simple story. He cannot say that his metrics for

[1] This is not intended as a serious hypothesis about the actual genesis of Grünbaum's ideas.

discrete space tell an intrinsic story of granular lengths. If there can be an intrinsic story about the length of grain-intervals there can be a similar intrinsic story about the length of continuous intervals, too. So Grünbaum's intrinsic measure story has a quite different plot. It is not about size and length of grains but about sets – how many members they have, which ones they are, or how many grains lie between them and an edge of space. The fatal snag for this tale is simply that it is about the wrong heroes.

9 Intrinsic absolute spatial metric

1 Discovering a metric: units and standards[1]

Now let us try a more constructive task. We want to understand how a space can have a metric that is intrinsic and unique. If it is both intrinsic and unique, then we will call the metric absolute. I am not suggesting that we can prove that the metric must be absolute. I am not sure what a proof of that would be like. I simply aim to make it seem plausible and intuitively appealing that the metric is not a convention made by us but a structural feature of space itself. A great deal of the hard work needed for this is already behind us. We have finished our rather long critique of conventionalism generally and of Grünbaum's metrical conventionalism in particular. There are now three things to do. *First* we have to separate the question how to find out about metrical structure from the question what has the structure we discover. *Second*, we need to understand in just what way metrical ideas are primitive. *Lastly*, we need to relate to one another the different strata within metric structure. Affine and metric geometries add further axioms to purely topological ones. We will bear this in mind and talk about each of these three structures in turn. The topology and affinity contain the deeper and more general elements of structure. The first two tasks are fairly simple, so let us take them in the order just given.

From here on I will take it that space is continuous. There is no need to bring discrete or countable spaces into the picture again.

It is pretty obvious that we find out about space's metric by tinkering with things in the space though they are not parts of it. We can defend the idea that space has an intrinsic metric (and just one) without denying the obvious. It is usual to tell that two spatial intervals are the same length by using the same measuring rod on them both. We carry the rod about and assume that it stays the same length as it travels. I see this assumption as a factual one. Now, the rod is in space, but it is not a bit of space: we have got into the habit of saying that this means the rod is

---

[1] In this chapter, some notation from differential geometry is introduced and explained. But what is philosophically relevant does not depend on the notation, though it is very useful and compact. Paragraphs are arranged in §§ 3–6 and 8 so that those which explain notation are easily identified and may be skipped without loss of continuity; §7 is not presupposed in §8 or later.

an extrinsic thing. I urge the view that we use the rod, an extrinsic thing, as a way of getting evidence for a theory about the metric of space. There is no need to withhold metrical facts from the intervals themselves, just because the rod is only in space but not part of it. Nothing stops the theory from being about equal or unequal intervals. I am simply marking off the epistemology of the theory from the theory itself.

Let us look (but not, yet, go) a bit further than this, even if it stops being obvious. We have looked at objects as utterly extrinsic to space, lying inertly in it while it passively contains them. There is no necessity about this picture. It is compelling only if we think space is The Void (a non-entity), perhaps because we think no one can frame for himself the concept of a thing which is unobservable, yet pervasive and macroscopic. But Newton affronted the philosophical conservatism of his day by tying the behaviour of force-free bodies tight to the metrics of space and time: the body moves on a linear path across equal distances in equal times, or stays at rest. The strength of these ideas in dynamics carried the field against conservatism, at least in physics. I think a profound extension of these ideas of Newton really lies behind the remarks of Riemann and Weyl, which Grünbaum takes as expressing his kind of conventionalism. In General Relativity, spacetime and things interact in a much deeper way than ever before, more coherently and intelligibly, too. Further, it powerfully suggests a picture in which things are parts of spacetime, after all, and in no way extrinsic. But even if we think just that extrinsic things (mass energy) mould the metric of spacetime causally it does not follow that the metric is not intrinsic. The fact that water can be made hot by an (extrinsic) electric element does not mean that the heat is not really in the water, after all. However, let's stick, for now, to more obvious ideas in which things and space do not interact, so that wielding things merely tells us about the size and shape of space. We can look at more intimate relations between them later.

Finding out about the metric is a much more complex, indirect business than the familiar stories about the standard metre-bar in Paris would suggest. The metric is never fixed by a coordinative definition or a stipulation that some rod or other will always count as the same in length whenever and wherever we use it. This is easily missed. The fact that picking a unit is obviously conventional and arbitrary masks it. When we are fixing which length is to count as a metre we have to settle how far apart to put the scratches on the platinum–iridium bar. Let us call the problem where to put the scratches the unit problem. Then the unit problem is hardly a problem at all. We could err in putting the scratches awkwardly close together or unmanageably far apart. But there is no reason why we should still not do something fairly silly (on the face of it, at least) like putting the scratches as far apart as the length of a king's foot. Choosing some fraction of the earth's circumference is more sensible. Again, how we chop the unit up into subunits – inches or centimetres – can be irksome for later calculators. It is, also, rather more practical if everyone uses the same unit. The metre will displace the yard and we can all use the same handy techniques of decimal calculating. What is silly or sensible in all this is obviously a matter of what is or isn't convenient. But though we do look to convenience in choosing a unit and we intend that it should make things simple for us, no criteria of convenience and simplicity uniquely determine a unit.

A far more important thing, in the general business of fixing lengths, is settling invariance. What really matters is not how far apart the scratches go, but that they go down in such a way that we can get that same length again, wherever and whenever we want it. Let us call this problem the standard-problem. It is a problem, all right, since it requires invariance. So we scratch a platinum–iridium bar, not a copper one and especially not a rubber band. The nature of the bar is what counts, the distance between the scratches does not. This is a matter of different hypotheses about the world – about things and space. Different systems of units do not reflect different hypotheses about the world. For it is part of our metrical theory about space that 1 metre = 1.1 yards (approx.); indeed, it is part of every metrical theory. To call the swimming pool 50 metres long is to say that it is 55 yards long. So the unit problem is not a problem since it is not a theory about things and space. But the problem of standards is harder. It does need a theory about space and things.

If two people judge differently about the equality of two intervals in space, then the only way to make their difference not a dispute is to cut out some aspect of what they seem to assert. (This is just what conventionalists do, of course.) There is no question of equating one full-blooded assertion with the other. As such they are conflicting hypotheses on any straightforward view. There is a dispute within a theory. Further, we can correct judgements based on the use of the platinum–iridium bar, by retreating to a measure of length which exploits the radiation of krypton, or whatever it might be. We dig deeper into the metrical theory. It is no tautology that the metre-bar in Paris is exactly a metre long (under specified conditions of temperature, etc.). The statement entails that the bar is always the same length (under those conditions) which it might very well not be, as a matter of fact. Unless we are to reel back into the abyss of conventionalism, we must accept that there is nothing trivial, arbitrary or conventional in the statement *as a judgement of invariance*. It is a statement within a factual theory. The judgement is as much about the spatial intervals the bar fills from time to time as it is about the bar itself. It says that the intervals filled by the bar are the same in length; it says so on the evidence that the bar does indeed fill one and did fill the other, and also that the bar has certain physical properties which, in our theory of things, we connect with invariance.

However, rods, bars, tapes and their transport really have very little to do conceptually with the metric of space. They are useful tools of measurement. We can carry them about and apply them easily. A good rod or tape gives satisfactory results over a wide range of distances and meets most needs for accuracy. Once upon a time, these were the only metric tools we had. But it was always wrong to argue, in the way so many philosophers and reflective scientists have, that the metric was fixed by a decree about some chosen object: that the platinum–iridium bar in Paris was *deemed* invariant in length. In Chapter 8, §6 we saw that the metric is really described by fundamental laws of physics and that it is naive to suppose that simple laying of rods end to end is the start and finish of the matter. As we refine our theory of the world, elaborate its laws and interlock them, a wide range of observations comes to bear, more or less directly, on the problem which spatial intervals equal which. The theory comes to specify new and more accurate ways of observing the metric, including ways which are more direct, if the theory is right, because they involve fewer extraneous assumptions. Physical measurement developed, in the beginning, from hypotheses that wooden rods, taut strings, the earth's crust and much else of the hardware of the primitive community, always gave pretty much the same length again. But wider, deeper, more systematic knowledge led us to criticise this without its rocking the philosophical boat. As we need to get more precise results, we begin to want a special rod used only under special conditions. This, in turn, can be given up in favour of very complex and expensive bits of apparatus, yielding very sensitive results by entirely novel techniques, say, from radiation sources. Finally, theory may reveal that these techniques rely on extraneous assumptions, about the constancy in time of some quantum relation, for example. Then we might look for a method which springs straight from the basic postulates of space or spacetime theory, as Marzke's light and particle interaction theory does, even though the instrumentation needed for this technique is still primitive.[1] In the light of these reflections, Riemann's remark that we must carry a magnitude forward, appears as no more than a broad metaphorical hint. We can certainly still take this hint, if we take it as meaning that we discover the metric by operating with things that are in the space but not of it. But we do not assume that things and space are utterly estranged – that space has no role in physics. On the contrary, as we look into the theory of spatial measurement we see from a new perspective that the metric is embedded deep in every branch of theory, that it is assumed in every prediction and confirmed in every observation, even if indirectly. These arguments hardly constitute a proof that metrical judgements are factual hypotheses about the size of space, but they certainly show at least that we cannot find a conventional basis for the metric in any simple coordinate definition.

Of course, there can be disagreements about the spatial facts. This can crop up not merely at the level of particular judgements of distance but at the level of competing global hypotheses about the metric; for example, whether curvature is everywhere the same or not. Given our somewhat loose sense of the word 'observation' the competing hypotheses will be consistent with much that is observed, some conceivable competitors being consistent with all that is or could be observed. But this does not mark off metrical hypotheses from other hypotheses in physics – those about electrons, for example. We posit objects, and hypotheses about them, from the beginning. Both material things and data of sense are posits of theories and these compete as to how neat and how soundly based they are. We aim for simple theories that are as sensitive to observation as we can make them. This policy has paid off. It has produced rich, stimulating, suggestive, powerful theories which are as general, intelligible, simple and deductively elegant as we can make them. We take these theories to be true because they have these features. As I said before, our epistemology appeals to coherence. The epistemology of the spatial metric relies on coherence perhaps with unusual heaviness. But we do take our metrical theory to be true. For us, truth is correspondence, so having reached our best theory, we take its comments on the metric to give us intrinsic factual information about the size of intervals.

[1] See Marzke and Wheeler in Chiu & Hoffman (1964), Chapter 3, pp. 46–8 and 62.

2 Determinates and determinables: the metric as a primitive

Let us take it, for now, that the naive view is correct after all. That is, that distances are absolute, that they measure quantities of space; that we find distance and measure roughly by looking, less roughly by using tapes or rods and finally by using quite complex apparatus. Newton held the 'naive' view, positing that various parts of space have size intrinsically, which belongs to them independently of things, though we use things to find out sizes. This is a factual hypothesis, but if it is primitive, as I suggested it is, how does it relate to the other primitive spatial hypotheses which we make in topology and other theories of space more general than metric geometry?

If metrical geometry is more explicit than topology then it is quite clear that the metric cannot be entailed by any description framed in the invariants of the more general language. To fix the local or global topology and to decide which curves are geodesics still leaves metrical questions largely open. It is true that if a space has a global topology of a certain sort, that may very well rule out some kinds of global metric. There can be no global Euclidean metric on spaces with spherical (S_3) topology. Further, even if the space has a Euclidean metric, that does not entail by itself which classes of intervals are equivalence classes. Any affine transformation of a Euclidean space (effected, say, by linear transformation of its coordinates) produces only another Euclidean space. But it will be another since congruences will not generally be preserved under these transformations even though geometry type is. This simply elaborates what was already said, that metrical sentences characteristically entail affine or topological sentences, but not vice versa. But it raises the question in what way metrical predicates are primitive, if there are necessary conditions for their being true of things (e.g. intervals) found in pre-metrical language.

I think this is a complex question; but I doubt that we need to probe it thoroughly for our purposes. I will content myself and I hope, the reader, with a model which shows off at least some important features of the matter. The model is that metric geometry is related to the more general and deeper geometries as a determinate is to a determinable,[1] not as a species to a genus. But what does that mean? Well, let us look at species and genus first. Husbands are a species of the genus men since a husband is a man who is married. In the formal mode 'x is a husband' means the same as the conjunction of 'x is a man' with 'x is married'.

[1] See Johnson (1964), Part 1, Chapter 11, § 4.

In terms of classes, the class of husband is the intersection of the class of men and the class of married things. But determinates and determinables are not related like this. The classical example is from colour. Though red is a colour, it is simply red, not a colour which is... In the formal mode, we can certainly agree that '*x* is coloured' is part of the meaning of '*x* is red'. But it is not a conjunctive part. There is no open sentence '*x* is *F*' such that '*x* is red' means the same as the conjunction of '*x* is coloured' with '*x* is *F*', unless we include the trivial case of the sentence '*x* is red' itself. Red things are just coloured things that are red and we can find nothing more analytical to say than that.

Something like this is true of geometry. We begin by saying what the space is like topologically, on both the local and global scales. Then, as will be shown in detail later, we simply single out a set of curves as geodesics (locally and globally). But this is not a matter of saying which further conditions these curves meet in order to be geodesics. Simply, we move from describing topological structure to describing affine structure by giving affine connexions in a primitive fashion. It is like describing a thing as coloured, then giving the colour as red. Geometry is more complex than colour. There are various equivalent ways of saying what affine connexion the space has. But I do not think this really alters the fact that we do give it in a primitive way. The move to the metric is like the move from '*x* is red' to '*x* is scarlet',[1] for those, too, are related as determinable and determinate. Given that we can identify (in the topology) two segments of distinct curves, then the assertion that they are geodesical is a primitive consequence of the affinity, and the claim that they have a definite ratio in length is a final primitive determination of the metric: final because there is no way to make things more determinate than that. (Once again, all these claims will be made good in detail shortly.) Nothing I have said about colour suggests that a thing can be red but no definite shade of red, such as scarlet; nor that a thing can be coloured without being some colour or other. Similarly, I am not suggesting that a true space can have topological structure, but not an affine or metrical one; nor that a space can have no metric but only an affinity. Of course, we can describe it topologically without describing it affinely or metrically. I am aware of no sound reason to doubt that true physical space is fully metrical; nevertheless, I have no proof either, that this is the case, so I will make no assumption either way.

Obviously, the state of logical affairs for geometry is far more complex than it is for colour (or for shape, which is also a determinable having

---

[1] As distinct from moving to some other shade of red, such as crimson.

determinates such as triangularity). But the analogy is suggestive and it sheds a little more light on why it has seemed so plausible for so long to take the metric as a convention. For, given the topology, the metric is neither fixed by it nor by further pre-metrical conditions. The analytical conscience rebels, perhaps, at the idea of a primitive relation, but not a purely sensory one, holding in a 'medium' which we cannot see, though we can see across it, and do see things in it. A metric which seems to 'come from nowhere' might seem to need a convention to support it. But this impression springs from a particular set of epistemological convictions which we saw reason to reject.

3 Basic structures: topology and coordinates

Let us take up the story of the metric from the point, at the end of Chapter 7, where we decided that topology is factual.[1] Suppose that the points and neighbourhoods of our physical space are topologically related to one another in some way. This topology is intrinsic to the space and depends on nothing but itself; it is simply the way space is. So it would be an absolute fact if space had the topology of E_3: space would be a continuous differentiable manifold, infinite and three-dimensional. Any closed surface in it would define one contractible and one non-contractible space. Maybe physical space is not like this, in fact, but like S_3 instead. But, however it is, whatever its topology, that is an absolute primitive fact about it.

Suppose space is at least locally Euclidean in topology: it can be broken up into regions so that there is a continuous 1–1 mapping of any region onto a region of E_3. This leaves room for a wide range of global topological structures, but in each case we can construct a coordinate system for the space, however it is made up over all. The aim of a system of coordinates, at this stage, is simply to name points in a useful way. Numerals are very useful as names, since we can make use of their easily understood orderings to reflect the ordering of the points. Co-ordinate numbers are not yet describing any quantitative relations. The coordinate system can be understood as a system of curves. In any region the coordinate curves are arbitrary save for forming three families, so that no two lines in the same family intersect. The families each cover the region. All of this means that each point of the space lies on the

[1] For a quite different way of approaching the metric from the topology, see A. Fine's very lucid account (1971).

intersection of three lines, one from each family. The points are assigned number triples in a continuous fashion; that is, each coordinate curve can be described by keeping two coordinates constant and letting the third vary continuously. Since the curves are selected arbitrarily and the numbers assigned in any continuous way, the coordinates have no metrical significance. Even so, for three points on a coordinate curve, one between two others, the coordinate difference Δx_k (k = the variable coordinate) is obviously smaller from the point between to either of the others, than the difference between the outer points.

We need some standard ways of writing coordinate numbers down, briefly and simply. An obvious way is to write $P(x_1, x_2, x_3)$[1] for the coordinates of point P. But it is shorter and quite usual to denote any point by the expression '$\{x_i\}$' which means the set of its three coordinates x_1, x_2, x_3 (i = 1, 2, 3). This merely labels or tickets the points.

Given a coordinate system of this sort for any region, we can make coordinates for the whole space by an atlas of such systems for various regions. Within these regions and also where they overlap, we assume that we can transform from one system to the other in the way to be described next.

These rather loose prescriptions mean that we can construct coordinate systems in many (infinitely many) ways. However it is done, there will obviously be some smooth function from the coordinates of a point in one system to its coordinates in another. Any one coordinate of a point P in a new system S' will be a function of all the coordinates of P in the old system S and vice versa (Lieber (1936), Chapter 15). It is assumed that we are dealing with a three-space, though all results generalise to n-spaces unless the contrary is explicitly signalled. In actual practice, we may break the rule that only a smooth function can give one coordinate system from another. In E_3, polar coordinates cannot be reached by any continuous transformation of Cartesian coordinates. But the rules still allow us great freedom so we will stick to them in our purely theoretical thinking.

We can write this out more formally, in compact style, by making the obvious requirement that the coordinates $\{x'_i\}$ of a point P in a new coordinate system S' be a continuous differentiable function of the coordinates $\{x_i\}$ in the old system S. We assume an inverse for this function. The relations which take us from S to S' or back can be

---

[1] For a spacetime point-event, we can write $P(x_0, x_1, x_2, x_3)$ where x_0 is the time coordinate.

written out in any of the following ways

$$x'_i = f(x_1, x_2, x_3); \quad x_i = f^{-1}(x'_1, x'_2, x'_3)$$
$$x'_i = f(\{x_j\}); \quad x_i = f^{-1}(\{x'_j\})$$
$$x'_i = x'_i(\{x_j\}); \quad x_i = x_i(\{x'_j\})$$

It is quite common to omit the braces which indicate that x_i is a function of the set of dashed coordinates and at times to use Greek sub- (or super-)scripts to replace the index indicator 'i'. So the above can be expressed very briefly as:

$$x'_i = x'_i(x_j); \quad x_i = x_i(x'_j)$$

There would be no value in a ticket for each point, so that for any change of coordinate systems the transformation function will find it again, unless we foresaw something interesting to say ultimately about the points. As Schrödinger puts it 'Our list of labels would amount to a list of (grammatical) subjects without predicates; or to writing out an elaborate list of addresses without any intention even to bother who or what is to be found at these addresses' (Schrödinger (1963), p. 4). This is not to say that the points are different from one another in themselves. They are indistinguishable, each taken by itself. To give some property to a point is really to say something about the space around it. If the property varies continuously through the space, as we saw curvature does in the case of surfaces (Chapter 4), then we want to specify the property *at each point*. This means attaching a number to the point which belongs to it in virtue of the geometric properties of the limiting region round it. The new number will be like a predicate which is true of some point or other, whereas the coordinate numbers are simply an arbitrary, non-descriptive name of the point. Topological structures alone seldom fix a number for any point, but unless we envisage the point as having some character in this way, the elaborate systems of coordinates are idle. We need to develop the idea of purely geometric features which we can look on as characterising the space at each point. We will use these to arrive at metrical structure. What form will our description of them take?

4 Tensors and vectors: features of space

In the case of surfaces, we can characterise curvature, for example, by a number which is called a scalar. To call a number a scalar means that it characterises the space[1] at the point in such a way as to be independent of the coordinates. If the quantity is defined by a scalar at each point of the space, we have a scalar field. A scalar is invariant under any transformation of the coordinates: so is a scalar field. But not every number attached to a point is a scalar: coordinate numbers themselves are plainly not, but a less trivial example is the speed of a particle at a given time. Speed is always speed *in some coordinate system.*

Consider the two-space of 'sea level' on the earth. This is a somewhat abstract idea, since the surface at sea level exists at a place whether there is sea there or not, or whether the ocean surface is rough or smooth. We could assign a number to every point on the earth's surface which would indicate how far the solid earth surface is from sea level at the point, a minus quantity sign giving the depth below sea level. This number would be a scalar. Since the variation in height is (mostly) smooth from point to point, the scalars over the whole sea level surface form a scalar field. We generally use a familiar latitude and longitude coordinate grid to label terrestrial points, so the scalar field could be given by ascribing the scalar number as a kind of predicate to the coordinate numbers in the usual grid system. If we were to change the grid system we would change the coordinate numbers of places like London, Wichita, Mt Everest and Adelaide. We would not change the height above or below sea level, of course, so the scalar number at the point is not changed. It is an invariant under coordinate transformation. Now a scalar is only the simplest example of an ordered set of numbers which may characterise a point in space for some purpose or other. Tensors are particularly important ordered sets of this kind, and they include scalars. The numbers in a set that is a tensor are its *components* and each of them is related to a coordinate of the space. That is, we use the coordinates to provide means for describing some entity in the space, though the entity itself in no way depends on the coordinates. The components are not scalars: they change under coordinate transformation. But they change in such a way that we might almost regard the whole tensor as a kind of invariant. The point is that the tensor transforms *as the coordinates do*; that is how the new components manage to describe the same thing. Sometimes tensors are in fact called invariants for this reason.[2] We can regard tensors as the basic form in which we will cast point by point descriptions of space. We will reach metrical structure by way of tensors.

[1] Or, perhaps, some property of a body or gas at the point, such as temperature.
[2] Lieber (1936), Chapter 15.

After scalars, the next simplest example of a tensor is a vector. Think of a vector as a kind of arrow. It has quantity (length) and direction. Any force is a vector. The force pushes at a point in a certain direction and with a certain magnitude. Another very easy example of a vector in metrical three-space is a step in the space from one point to another. It is the line at a point which joins it to some other point and which has a length in a definite direction. Take a point P and a small arrow from P to a nearby point Q. We can characterise the length and direction of the arrow from P to Q by a pair of numbers in two-space, a triple in three-space and so on. The coordinate system is used in fixing which numbers to include in the pair. We simply give the coordinate differences between P and Q. The signs tell us direction. The components of the vector from P to Q differ just in sign from the components of the vector from Q to P.

Since we rely on the coordinates to describe the step in space, our description changes when we change coordinates. But the step (or the force) is the same: the electric field is just as before; Edinburgh is as far away from London in the same direction whether we use latitude and longitude differences to express this, or some other means. Q is just the same distance from P as before and along the same track. Thus, though the numbers change when we change coordinate systems, what is being described at the point does not. Further, when we change systems, the change in vector components is intimately related to the coordinate transformation function, which is predictable enough. Indeed, vectors can be defined by the law for their transformation and so can tensors more generally. We will look at some notation to express these ideas more precisely and compactly in a moment. But, first, an example of a full-blooded tensor.

Probably the easiest tensor to understand visually is the tensor of stress in an elastic solid, such as a block of soft rubber.[1] Imagine various forces impinging on its surface. Then there are stresses at all the points inside the block. We could see this if we could cut the block along a plane. The material on either side of the cut would bulge or shrink (in general) revealing that the matter on the other side was exerting forces to keep both surfaces plane when they were together. Suppose that we don't actually cut the block along this plane, but use it to express the forces at work in the block. Consider a very small area of this plane, so that the magnitude of the force is proportional to the area.

Choosing this small area in the plane does not fix the direction of the

[1] Feynman *et al.* (1963), Vol. 2, Chapter 31, especially pp. 31–6; Lieber (1936), Chapter 14.

force acting on it. It may or may not act in a direction perpendicular to the area. If we change the direction of the plane through the point we are interested in, we will not get the same force as before. Evidently, the stress is a complex matter. We need one vector to express the force and another to express the size and orientation of the surface in question. We need a tensor of rank two (roughly, a complex of two vectors). This will have 3^2 components in three-space, n^2 in n-space. Tensors may have arbitrarily high rank. A tensor of rank four in three-space has 3^4 components; a tensor of rank m in n-space has n^m components.

Analytically, the story looks like this. Our function for transforming the coordinates makes any one coordinate in S' a continuous differentiable function of all three coordinates in the S system. Tensors characteristically lead us to deal with the limiting regions round points so when we think about how to find them again in a new system we will look to how the coordinate differentials dx_i transform into the dx'_i. They transform according to the partial differential equations:

$$dx'_i = \frac{\partial x'_i}{\partial x_k} \, dx_k. \tag{1}$$

The repeated index k, on the right, shows that the term has to be expanded as a sum of terms of that form for each element in the index set $k = 1, 2, 3$. So (1) abbreviates

$$dx'_i = \frac{\partial x'_i}{\partial x_1} \, dx_1 + \frac{\partial x'_i}{\partial x_2} \, dx_2 + \frac{\partial x'_i}{\partial x_3} \, dx_3 \tag{2}$$

and (2) abbreviates three equations, one for each member in the index set $i = \{1, 2, 3\}$. The terms $\partial x'_i/\partial x_k$ express the rate at which an S' coordinate changes for a small change in one coordinate of S, the other coordinates of S being held constant. So the expression $\partial x'_2/\partial x_3$ for example, represents the change in x'_2 per unit change in x_3 only. So it is the partial derivative of the function. This 'ratio'[1] is then multiplied by the total change in x_3, that is, dx_3, and the partial changes are all summed as in (2), to get the total x'_2 change, dx'_2. We will transform the tensor components in just the same way as we can see if we reflect on the simpler tensors (see Lieber (1936), Chapter 15).

A vector is a set of three numbers, called its components, which are linked each to one coordinate. We can express a step or separation vector by the coordinate differences between the points, Δx_1, Δx_2, Δx_3; these are its components. We can express this same vector in a new coordinate

---

[1] It is, strictly, the limit of a series of ratios, of course.

system by changing its components in the same way as we change the coordinates themselves. The new coordinate differences $\Delta x'_i$ make up an ordered set identifying the same step in space at the point. Not all vectors are steps in space, but we will call any ordered set of three numbers a vector on condition that it transforms like the coordinates; that is, as follows:

$$A'^i = \frac{\partial x'_i}{\partial x_k} A^k.$$

But we are running rather ahead of ourselves. These ideas are all at home in metrical geometry, but we were busy trying to see how much could be said in topological language by way of foundations for a metric. Now the most revealing path to strike out on, if we want to travel from topology to metric, is one that leads from proto vectors and tensors to the real metrical thing. So we need to see what we can find in topological structure which is vector- or tensor-like. Let's begin by looking for something in a metrical space which will be there for all continuous 1–1 mappings of the space into itself, and thus is also a topological invariant. Then we can look at the same thing all over again from a purely topological perspective.

5 Directions at points: vectors in topology

In E_3, take any family of continuous curves which all intersect on just one point, but which cover the space throughout some finite region round the point. All such families of curves have something in common. Any curve through the point has a tangent at it which can be reached by considering the series of shorter and shorter segments containing the point – by the limiting operation of differentiation. So any family of curves of the kind described also defines the set of tangents at the point. This set of tangents is common to all families of curves through the point, including the pencil of straight lines. Since straight lines are such simple curves, we can regard all families of curves through a point as approaching in the limit the pencil of lines through the point. The set, or the pencil, of tangents at the point can properly be called the set of 'line elements' at the point. Any *directed* line element at the point is a vector, which is called a tangent vector, partly to remind us how we came by it. We can regard the tangent vector as expressed by the triple of coordinate differentials, of the form dx_i.

Clearly enough this can be generalised. We put only topological restrictions on the curves through the point and found the tangent vectors common to all families. So this pencil of small vectors must be there in all topological transformations of the space which map it into a space composed of finite regions each topologically equivalent to a region of E_3. These are just the cases that concern us. From our broader viewpoint, the tangents are no longer quite infinitesimal directed segments of *Euclidean straight lines*, though the expression 'line element' is proper enough as we shall see when the vectors generate geodesics. The tangent vectors are purely spatial things. They are parts of the space as the curves themselves are. They exist no matter which of the various metrical structures a space of this topological kind may turn out to have.

Let me stress again that coordinates and coordinate expressions are not of the essence in discussing vectors, including tangent vectors. They are convenient but dispensable. The vector is like a little arrow embedded in space and an integral part of it. The coordinates lie on the space like cobwebs, so to speak, to be blown away by the next whim of nomenclature. For other approaches to tangent vectors, including a coordinate free approach see Misner *et al.* (1973), Chapter 9 and Schrödinger (1963), Chapter 1.

What are these tangent vectors, if we think about them in a purely topological way? They are directions round the point. Again, they are purely spatial things – *of* the space, not just *in* it. The argument was that the set of directions at any point is well defined in topology. So we can speak of vectors and tensors at points in space on that minimal purely geometrical basis. But this does not mean that directions link up from point to point as a matter of topology. Nothing, therefore, yet says how vectors and tensors generally compare at different points either. It is easy to see that this may not be possible at all in the large scale, for some metric spaces. Take the sphere, for example (see fig. 35). Place two arrows in the surface at the 'equator', both pointing 'north'. Perhaps they have the same direction? Suppose they are a quadrant's distance apart. Move them along their own length (in their own directions) till they meet at the pole. They will be at right angles to one another and not pointing in the same direction at all. But there were no other vectors, in their original positions, which were better candidates for having the same direction. We may not be able to speak of the same direction at distant points.   So a topological vector-pencil is a set of all directions, but only at one point. We need a few more observations on tensors, then we can take the next step in structure and compare directions at nearby points.

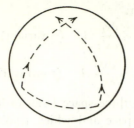

Fig. 35

Vectors and tensors may be covariant or contravariant. What does this mean? We said that a vector must transform as the coordinates do. That shows that it is in the space and not just in a particular coordinate labelling. Our specimen vector had the coordinate differentials as its components and transformed as they did, naturally enough. But that is only one kind of vector, or rather one way of expressing a vector in components. If we consider a coordinate system in metric space as itself based on a set of *unit vectors* at the point,[1] then we can express any vector at the point in terms of products with these base vectors in two ways (which need not hold us up now).[2] In one form the vector components vary under transformation as the *base vectors*; in the other form the components vary in a contrary way. So the vector, as an ordered set of three numbers, may be a covariant or a contravariant vector. When we express the line element as a vector by means of the coordinate differentials, we express it as a contravariant vector. Tensors are, again, somewhat more complex, having indices of covariance or contravariance. They may be mixed or pure (either pure co- or contra-variant).

In writing out the partial differential transformation equations, the 'partial ratio' for a covariant vector is the inverse of the ratio for a contravariant vector. The convention is that the index for a covariant vector appears as a subscript to the vector (as the primed differential is the 'denominator' of the transformation ratio) whereas the index is written as a superscript for contravariant vectors (the primed differential is the 'numerator' of the transformation ratio). Thus we have two transformation laws each characterising a kind of vector, the first called covariant, the second contravariant. The coordinate differentials

---

[1] See Aleksandrov *et al.* (1964), Chapter 16, §2.
[2] See Adler *et al.* (1965), pp. 32–8 for a clear and complete account of this. For Cartesian coordinates the distinction vanishes, but this already requires Euclidean space. See Misner *et al.* (1973), Chapter 2, §5 and Chapter 4, §3 for applications.

are a contravariant vector.

$$A'_i = \frac{\partial x_k}{\partial x'_i} A_k \quad \text{(covariant)}$$

$$A'^i = \frac{\partial x'_i}{\partial x_k} A^k \quad \text{(contravariant)}$$

A tensor of rank n in three-space is an ordered set of 3^n numbers assigned to a point of space and which transform in a certain way. A tensor of rank n will have m covariant indices and p contravariant indices, its general form being $A_{k...l}^{i...j}$ where $k...l$ are the n covariant indices and $i...j$ are the p contravariant indices $(m+p = n)$. The transformation law for a tensor, which defines the tensor formally, in our premetrical sense, is

$$A'^{i...j}_{k...l} = \frac{\partial x'_i}{\partial x_d} \cdots \frac{\partial x'_j}{\partial x_f} \cdot \frac{\partial x_g}{\partial x'_k} \cdots \frac{\partial x_h}{\partial x'_l} A_{g...h}^{d...f}.$$

Tensors (vectors and scalars) each hold *at a point*. Given two vectors or tensors at the same point we can say they are equal if their components are equal, otherwise they are unequal. We can add them, form their products (there are scalar and tensor products of tensors)[1] and develop an elegant algebra for them, provided we do the adding and multiplying at the same point. But it would be useful to be able to do more. The idea of a tensor field and of how the field changes from point to point, locally anyhow, still has to be given a foundation. Comparing directions at neighbouring points, which is the next step in structure, will let us make sense of operations on a tensor field.

One reason for stressing the fact that a vector, for example, is always a tensor at a point is that this is so easily overlooked in Euclidean geometry. If we use Cartesian coordinates, then the same components define at any point a vector of constant magnitude and direction. So, unless we have a vector field changing from point to point (as occurs in physics with vector analysis of the electromagnetic field, for example) it is unnecessary to specify the point at which we set the vector up or give its components. Even if the vector field is seen as changing from point to point, vector addition is apt to be learnt geometrically by adding two vectors at P like arrows, 'nose to tail', where it is assumed to be quite unproblematic how to construct at the 'nose' of **u** the vector **v** even though **v** is strictly defined only for the same point as **u** (its 'tail'), not at the point to which **u** extends (its 'nose'). But the problem of adding

[1] Adler *et al.* (1965), Chapter 1; Lieber (1936), Chapter 17.

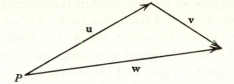

Fig. 36. **w** is the Euclidean vector at *P* which is the sum of the vectors **u** and **v**, both at *P*.

vectors (and tensors) in spaces more generally is harder as we shall see.

How far have we got? At present we have described space – a range of spaces, in fact – by specifying that their topology is locally Euclidean, which means that they can be covered by a coordinate system (or patchwork of systems) of the sort we specified. Even at the level of this very general circle of ideas we could describe an object in the space which is a (contravariant) vector: the line element at a point expressed by the three coordinate differentials dx_i. That was the main point of this section. But I also defined covariant vectors and tensors generally at this premetrical level, without giving any examples, however. Vectors and tensors are defined here by the way their components transform, rather than by being directed quantities. Since the vector-pencil of line elements is a part of space (or a complex of parts) the structure we focused on is intrinsic to the space.

6 *Affine structure: neighbouring directions*

Grant that the set of directions round any point in physical space is intrinsic to it. Now I want to suggest that some more structure is intrinsic, too. The structure is this: take the pencil of line elements at some point *P* and the nearby pencil of line elements at *Q*. These were shown to be of the space; intrinsic to it at the points. Then this is surely intrinsic, too, that if we choose a line element from the pencil at *P* there is a definite spine at *Q* which has the same direction. Which spine it is depends just on the small displacement dx_i which is the direction of *Q* from *P*. That is, it is intrinsic to the space that it has enough structure or shape to align directions at any point with those at a nearby point. This means that we can line up neighbouring pencils one with another. But this does not mean that we can talk about directions globally in the space. The intrinsic pencil line up is a local, neighbourhood affair. Whether it adds up (integrates) to global directions, as it does in Euclidean space, or whether it is only local (non-integrable as it is on the sphere), is a matter for further structure (including global topology). This local alignment is called the 'affine connexion' or the 'affinity' of the space. Our hypothesis is that this is intuitively (plausibly) an intrinsic connexion in any physical space since we are not talking about anything other than the space and its points. It is a primitive, non-topological structure.[1]

The affinity lets us 'transplant' any vector from P to nearby Q. It allows parallel 'transport' of vectors.[2] It means that given any vector components at P we can find from the affinity which components at Q define the same vector. So we can also tell whether the vector changes or not from P to Q given the affinity and the components at P and at Q. (Whether the same vector has the same components at P and at Q depends not only on the space but also on the coordinate system. That is obvious enough from the possibility of curvilinear coordinates in Euclidean space.) Once vectors at P compare with vectors at Q we can begin to use the idea of a tensor field. We can say whether the field is constant or changes from point to point; we can differentiate it and open the way for some powerful mathematical machinery. Affine structure offers a toehold for a great deal of geometrical (and physical) theory.

Given that the affine structure is intrinsic and (equally plausibly) unique, then it follows that a curve's being geodesical or not is intrinsic and unique – absolute. The affinity determines geodesics. Put graphically, it works like this. Start from P and choose a direction (displacement vector) dx_i from the vector-pencil. Follow it to nearby Q and pick the same spine (direction, or displacement vector) from the pencil at Q which will lead you to the pencil at R where you pick that spine again. How we select the spines at Q and R is fixed by the original direction at P together with the affinity. The path traced out by this process is a geodesic. We slide the little vector along itself always, so the path marked is geodesical: a kind of uniform motion (Weyl (1922), pp. 114–16). This is more graphic than precise, of course. The curve is defined (as the dx_i are) by the limit of a series. The members of the series are broken 'dog leg curves' made up of vector 'steps' from point to point, as in fig. 37. The series is generated by taking the vector 'steps' smaller and smaller. The process even permits us a kind of primitive comparison of

---

[1] Adler (1958), Chapter 2, §1; Bergmann (1968), pp. 67–74; Schrödinger (1963), Chapter 5; Weyl (1922), §14.

[2] The vectors are not literally transplanted or transported, of course. The pencil of line elements is a spatial entity and cannot intelligibly be supposed to move. 'Transport' is a metaphor for comparison (including equality) at nearby places.

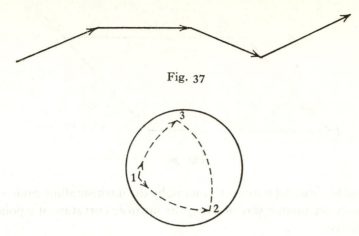

Fig. 37

Fig. 38

length of two disjoint (not overlapping) intervals on the same geodesic. We count the number of 'steps' in each, or rather we find the ratio of the number of steps in each as the steps go to zero (Schrödinger (1963), Chapter 7, especially p. 53). But this does not give us a way to compare length along two different geodesics.

Now we come upon a weighty fact: the affinity determines a fundamental curvature of the space, point by point. The curvature is intrinsic to the space if the affinity is. The easiest way to see how this comes about is to take any vector in the surface and carry it round a curve as the affine connexion tells us how to do, from point to point. (Notice that we will not be sliding the vector along itself, since we take it round a curve.) Then we will see how much it has changed *in direction* when it 'comes back'. The Gaussian curvature at the point is the limit of the ratio of this angle to the area bounded by the curve as the area gets vanishingly small. We can get a picture of how this works (involving a bit more structure admittedly) by looking at the sphere again. Take a vector on a sphere and regard the curve which it points along as equator to another point (3 in fig. 38) which is a pole. Slide the vector along its length to a point a quadrant's distance from the starting point (from 1 to 2 in the figure). Now carry it up to the pole (from 2 to 3) keeping it at a constant angle to this geodesical path. Finally bring it back to 1 (sliding it 'backwards' along its own length). The vector will be at right angles to its original direction, though at each point its 'transport' was parallel – either along itself or at a constant angle to another geodesic. The final angle (and the area) point to a non-zero curvature which is

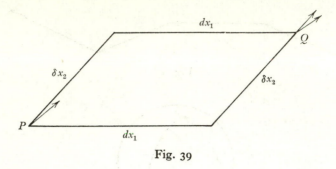

Fig. 39

betrayed by 'parallel transport', that is, by the intrinsic affine connexion. So here is yet another way of fixing the intrinsic curvature at a point of any surface.

For n-dimensional spaces quite generally, the curvature of the space at a point P is fixed by the curvatures of surfaces that contain P. The affinity carries the load, but by a technique rather like the one we used in fig. 35 (p. 201), though it is equivalent to the one just shown. A vector is given a small displacement in one coordinate direction, say x_1, then a further small displacement in another coordinate direction, say x_2. Then the process gets reversed. Taking the vector at P again, we give it a small displacement, first in the x_2 direction, then in the x_1 direction. This amounts to taking a small parallelogram in the x_1-x_2 coordinate surface, transporting **v** first round one pair of sides, then round the other and comparing the results at Q (see fig. 39). If this is done with the same vector for all the coordinates surfaces we can build up a complex measure of curvature, which is the curvature tensor at P (Weyl (1922), pp. 107–8).

7 An analytical picture of affine structure

We explained our claim that space is a particular by the claim that it has a shape. This turned out to mean that it has a curvature everywhere. So this idea is very important for the credibility and clearness of our main goal. Let us look at the question a little more analytically. In fact, this will mean no more than some useful notation. We are not going to calculate in the machinery of tensor analysis. This skirmish into analytical territory is illuminating, I think, but not vital for understanding the affinity as a structure intrinsic to space.

Let us go back to the first idea of the affinity, the idea that there is an intrinsic structure to the space whereby a vector at P can be compared with a nearby vector at Q as to whether it has changed or not. In Euclidean space, given only Cartesian coordinates, equal vectors at different points have equal components. But this cannot be the right way to define the identical vector at different points in general. Obviously this method is relative to a general style of fixing coordinates, viz. Cartesian ones, and breaks down, even in E_3, for general curvilinear coordinates. Since Cartesian coordinates are possible only in Euclidean space, it is a recipe which could not work outside these spaces. Vectors should compare at points in a way which is intrinsic to the space and is not dependent on the chance system of labelling. So the problem is to discover which changes in the components of some vector A^i (at P) give the same vector at Q; alternatively, how to tell from comparing the components of a vector A^i (at P) and B^i (at Q) how the vector at Q differs from the vector at P. The vector at Q which is identical with A^i (at P) can be expressed as the A^i minus some quantities δA^i, where δA^i depends both on A^i and the displacement vector (direction) dx_i from P to Q.

$$A^i \text{ (at } Q) = A^i \text{ (at } P) - \delta A^i$$

There are some obvious formal requirements on δA^i: it must allow A^i to transform as a vector both at P and at Q; it must vanish when either the dx_i do or when A^i at P does (when A^i is a null vector). The simplest form which meets the conditions makes the δA^i linearly dependent on the vector at P and on the direction from P to nearby Q. So we get the following bilinear form:

$$\delta A^i = -\Gamma^i_{kl} A^k dx_l$$

where Γ^i_{kl} is a three-indexed array of 3^3 (3^n for n-spaces) numbers. This array of numbers, to be fixed at each point of the space, describes the affinity for each point.

Given the Γ^i_{kl} for one coordinate system, we will naturally expect to be able to find it again when the coordinates are changed since it is an intrinsic feature of the space. This is indeed so, though the affinity does not transform as a tensor, since it also depends from point to point on coordinate differences.[1]

It is clear enough, from the form of these equations, that we can rely on the affinity, the Γ^i_{kl} at each point, to discover what we mean by the

---

[1] Schrödinger (1963), Chapter 5, especially pp. 39–40; Adler (1958), Chapter 2, §1; Bergmann (1968), pp. 61–74; Weyl (1922), §14.

differentiation of a vector (or tensor) field; more strictly, it lets us describe the covariant derivative of a vector field. That is, we will be able to say what it is for the vector field to change from point to point. Ordinary differentiation of a vector field tacitly assumes that the vector at P is unchanged at Q *if its components are unchanged.* It is Cartesian, Euclidean differentiation. This does not give us the concept of a derivative independent of coordinates and of a particular kind of space. We need to express a derivative of a tensor at a point so that the derivative is itself a tensor. The affinity tells us just how to take the difference in the vector at nearby points in a way that lets the derivative transform covariantly. So we get the following form of derivative

$$A_{i;j} = A_{i,j} - \Gamma^k{}_{ij} A_k$$

where the comma expresses ordinary, and the semicolon expresses covariant, differentiation. This tells us how the field is changing at the point.

To get the curvature we make use of this equation for covariant differentiation, and we differentiate twice in accordance with it. The usual way is the following: take a covariant vector A_i, differentiate it covariantly with respect to some coordinate k, then with respect to another, l. This is done by appropriate expansions of the equation above. This gives us on the left a tensor of rank three, $A_{i;kl}$. Now form the covariant derivatives of the same vector in the reverse order $A_{i;lk}$. Then subtract these tensor equations. The new equation leaves us still with a tensor of rank three on each side. *Subtracting these derivatives is really a compendious analytical way of taking the vector in opposite directions half way round a small parallelogram in each coordinate surface,* as I described before pictorially, at the end of §6. The subtraction tells us how the vector differs in each case. Which coordinate surface we have depends on which of the values 1, 2 and 3 we substitute for the indices i, k, l. We wind up with an equation showing a tensor of rank three on the left and a rather complex object, by which we have come to multiply our original simple vector in the differentiation process.[1] In short we have:

$$A_{i;kl} - A_{i;lk} = \Psi A_j$$

where Ψ is some tensorial object or other. We should notice two things about Ψ: first, that it is the coefficient of a covariant vector and second, that the product of Ψ and A_j is equal to a tensor of rank three with all

---

[1] Note that the covariant derivative gives us a function of the original vector A_k on the right as well as its ordinary derivative.

indices covariant. The rules of tensor algebra require, then, that Ψ' is a tensor of rank four, with one contravariant and three covariant indices. The fourth contravariant index is cancelled, as it were, in the product with the covariant vector A_j, so that the whole right side is a tensor of rank three, and covariant even though Ψ' is rank four, and mixed. Ψ' is written $R^j{}_{ikl}$ and called the key tensor. We can express this tensor in terms of the affinity, suitably differentiated, since we envisage moving a vector in opposite directions half way round a small area. It looks like this:

$$R^i{}_{klm} = -\Gamma^i{}_{kl;m} + \Gamma^i{}_{km;l} + \Gamma^i{}_{al}\Gamma^a{}_{km} - \Gamma^i{}_{am}\Gamma^a{}_{kl}$$

Fortunately, we need not actually calculate here with this tortuous object.[1]

Just how tortuous it really is varies with the number of dimensions of space. Obviously, the number of equations, and the number of terms in each equation, rises sharply with increasing dimensions. But the number of independent components varies with the number of dimensions, too. For surfaces, everything that can be said about curvature at a point can be given just by a curvature scalar, as we saw. For three-spaces, the number of independent components lets us off with a tensor of rank two which is the contraction[2] of $R^i{}_{klm}$. But the mixed four-tensor $R^i{}_{klm}$ is necessary and sufficient for fixing the curvature of n-spaces generally for $n > 3$.[3]

The affinity, then, provides us with a great deal of intrinsic structure. We have parallel vectors (directions) in small regions, geodesics with a primitive comparison in length for segments on the same geodesic and, finally, curvature and consequently shape for the space. Plausibly, all this structure is intrinsic to the space and absolute. We have rested only on the idea that there is a pencil of the line elements at every point and that they are lined up with one another in each neighbourhood. Now we can turn to the last step, the metric itself.

8 Metrical structure

The metric of space is a conjunction of two features: one is the comparison of length among the various line elements in the pencil at a point;

[1] See also Lieber (1936), Chapter 23; Weyl (1922), §15, especially pp. 119–20; Bergmann (1968), pp. 167–8; Adler *et al.* (1965), Chapter 5, §2; Riemann (1929), Part 2, §§3, 4.

[2] Lieber (1936), Chapters 20, 25; Adler *et al.* (1965), Chapter 1, §6.

[3] Bergmann (1968), pp. 172–4.

the other is the connexion of this comparison at a point P with that of a neighbouring pencil at point Q. My claim is just what it was before – that these two features can properly be assumed intrinsic to the space and unique for it, so that the metric becomes absolute.

Essentially, the measure of length of the line elements in a pencil links up with the understanding of them as vectors. Two vectors will multiply in a way that forms their *scalar product*; this is a pure number, a quantity. We can also multiply a vector $\mathbf{v}$ by itself in the same way and take the number v^2 as a function of the length of a vector: the length of the vector is the square root of this number; $\mathbf{u}$ and $\mathbf{v}$ are equal if $u^2 = v^2$. But all this is aimed, first, at the line elements, the dx_i vectors, which we take as having each a length comparable with the lengths of others. Call this length ds. Then we assume that this length is some quadratic function of the dx_i. This is not the only conceivable form which relates ds to the dx_i, but it is a simple one. The spaces for which it holds are Riemannian spaces.

We can write this out, putting a g coefficient before each product of the dx_i, with suffixes on each coefficient to tell us which product it governs. The general form of the equation looks like this in two-space:

$$ds^2 = g_{11}dx_1{}^2 + g_{12}dx_1 . dx_2 + g_{21}dx_2 . dx_1 + g_{22}dx_2{}^2$$

For n dimensions generally (using summation on repeated indices as on p. 198), it can be written briefly as follows:

$$ds^2 = g_{ik}dx_i dx_k \quad (g_{ik} = g_{ki}).$$

The numbers g_{ik}, which are the coefficients of the dx_i products at the point, form a tensor at the point. These numbers tell us how to multiply the dx_i vectors together at the point so as to yield up the length. What they say depends both on the intrinsic nature of the space at the point and also on the kind of coordinates in use.[1] This tensor transforms accordingly to the appropriate tensor law when we change coordinate systems. It describes a feature intrinsic to the space at the point which lets us compare the lengths of different line elements ds in terms of the dx_i, each of which is now seen as a line element along a coordinate curve through the point. The g_{ik} form what is called the 'metric tensor'. It exists at each point of the space and fixes the length of line elements there in the way described.

---

[1] For a simple account of what kind of quantities the g_{ik} might be see Lieber (1936), Chapters 13 and 21. For Cartesian coordinates, they are simply $g_{11} = g_{22} = 1; g_{12} = g_{21} = 0$.

But for the whole space to count as a metric space, *each point must connect metrically* with the points in the region round it. It is plausible to assume that this holds in such a way that the metric connexion is a particular intrinsic determination of the affine connexion. Weyl writes ((1922), p. 124) that this 'may almost be called the *key note of infinitesimal geometry*, inasmuch as it leads the logic of geometry to a wonderfully harmonious conclusion'. This leaves many possibilities open but let us follow Riemann in assuming that in physical space a length at any point compares with a length at any other, independent of the path along which the vector is transplanted. So, if we carry a vector round a closed path in curved space, its direction may change but its length will not. More generally, this means that the scalar product of two vectors is constant whenever they are parallelly transplanted along a path (according to the $\Gamma^i{}_{kl}$). Thus the $\Gamma^i{}_{kl}$ and the g_{ik} may be expressed in terms of one another like this:

$$\Gamma^i{}_{kl} = -\tfrac{1}{2}g^{ji}\left(\frac{\partial g_{kj}}{\partial x_l} + \frac{\partial g_{jl}}{\partial x_k} - \frac{\partial g_{lk}}{\partial x_j}\right),$$

which expressed the Γ's in terms of the *partial derivatives of the g*'s.[1] Just what this means need not hold us up. The curvature now links up with the metric by benefit of the earlier expression for curvature in terms of the Γ's (p. 209). In this form (and with its contravariant index lowered)[2] the tensor is known as the Riemann–Christoffel curvature tensor R_{ijkl}.

This completes the account of the various levels of structure that go to make up the metric of space. I call the structure intrinsic, because we were able to deal with purely geometrical entities throughout, once we put aside the question how to find out the structures of space. I do not know of any way to prove that conventions are not mixed up with the structure nor any way to prove that it is indeed primitive and absolute. I can see no reason at all to doubt that it is absolute. The burden of proof lies on those who wish to argue that the world is really less structured than theory says it is. We have looked into arguments along these lines, but did not find them good. So let us conclude that space has a definite real intrinsic structure in its metric, affinity and topology. This means that it has a shape and size in a way that I have tried at length to make clear. It shows just how space is a particular thing. That

[1] Weyl (1922), pp. 125–6; Adler *et al.* (1965), Chapter 2, §2; Bergmann (1968), pp. 71–3; Lieber (1936), Chapters 13 and 21.
[2] See Adler *et al.* (1965), Chapter 1, §8 on raising and lowering tensor indices.

is its place in our ontology, earned for it by the arguments of the book so far.

In the next chapter, we reflect on the problems and opportunities which facts about motion offer an absolutist theory about space and spacetime. Chapter 2 ended with the suggestion that an absolutist can explain handedness by appeal to the idea of a shape for space globally. This was not offered as a causal explanation but as a geometrical one. Whether the suggested explanation is an empty gesture depends really on whether or not geometric explanations are acceptable for physical phenomena in general. Beyond the traditional interest which philosophers have found in the question whether motion is relative or absolute, there lies a whole rich explanatory theory of dynamics explained geometrically in General Relativity. So the yield in Chapter 10 will be to expand and strengthen our early picture of the shape of space as an explanatory idea. We have just seen that affine connexions and metrics are basically matters about the shape of space (or spacetime) in small regions. In General Relativity it is exactly the local geometric structure of spacetime that carries the explanatory lead. To show this in detail would be an enormous task. I aim to present enough to underpin my philosophical conclusion.

10 The relativity of motion

1 Relativity as a philosopher's idea: motion as pure kinematics

In the history of Western thought, the chapter on the relativity of motion is long, extraordinary and unfinished. A vivid thread of philosophical dogmatism runs right through it. One remarkable combination of facts is that philosophers from Descartes to the present day have seen it as obvious, even as trivial, that motion is purely relative[1] whereas no theory of physics, with the doubtful exception of General Relativity, has been consistent with purely relative motion. It seems obvious that the concept of motion should be a very basic category of thought and, if not a primitive idea, then at least a very simple one. But the history of physics makes it clear that it is a rather intricate notion, inside physics at least. Does the open sentence 'x moves' express a cluster of concepts? This would be easier to accept were it not that the concepts of physics wind up inconsistent in a very strong sense with the ideas we might think of as the primitive ones. This is quite distinct from the better known, but still dramatic way in which physics has overthrown our ideas of space and time. My aim in this chapter is to point out the more significant features of the philosophical scene as it has changed from classical physics to Special and then to General Relativity.

But how does this interest tie in with the concerns of the book in general? No book about space can be complete without some discussion of the ancient question whether or not motion is relative. This is a huge topic. Lest it swamp the rest of the book, I will stick rather closely to those aspects of the problem which shed light on the idea of geometric explanation or which show its importance. At any point, we will never move far away from our concern with the shape of space or spacetime. I am eager to show that an understanding of rest and motion in the context of an increasingly geometrical style of explanation moves us in the direction of seeing rest and motion as absolute. There is a rather complex relation between the two ideas, which needs deeper exploration than I shall give it. But I hope that what I shall say may open a fresh perspective – to philosophers, at least – on this much-discussed issue.

Only the barest minimum of physics, consistent with providing a

[1] In a sense explained shortly.

basis which enables a reader to see how geometrical explanation takes over in General Relativity (GR) is given here. It is very tempting to think that a 'bare minimum' needs, in fact, a large slice of the foundations of this theory. It is profoundly beautiful and exciting, after all. I have given a fairly complete qualitative explanation, from scratch, of the kinematical ideas which are most striking and relevant in Special Relativity (SR). But my treatment of GR is fragmentary, by comparison. It is confined to one example (the clock paradox) and a brief explanation of the physical role of the ideas of affine and metrical geometries introduced in Chapter 9. I give frequent advice as to where the reader may turn for the more accessible, more detailed accounts. So let us press on.

It is not quite clear what you say if you maintain that motion is only relative. It sounds like a claim about the real semantic structure of sentences containing the verb 'to move'. It might be the claim that 'x moves' is satisfied by an object a if and only if a and some object b satisfy 'x moves relative to y' in the right order. But that is not what Newton, the arch-absolutist, denied. It was never his idea that the difference between moving and non-moving bodies is a qualitative one, constituted by non-relational attributes of things. By contrast, Leibniz, Newton's great contemporary opponent on the question of relativity, really did claim this.[1] That is part of Leibniz's metaphysics, not his physics, however. In fact, this semantic claim was never a bone of contention between them. Their battle was about two other statements: first, if motion is a relation, then it relates only familiar bodies to other bodies; second, if motion is a relation, then it is a symmetrical one. I will refer to the first as the materialist statement of relativity and to the second as the symmetry statement.

The relativity of motion, then, is a cluster of three statements: a semantic thesis, a materialist thesis and a symmetry thesis. Anyone is a relativist if he asserts most of them. At times, Locke, Berkeley, Mach, Poincaré, Einstein and Reichenbach asserted all three together. This characterisation is a bit rough, clearly. It gives only a standard picture and gives that rather vaguely. But it is good enough, I think, to start us along the road to discovering the fate of relativism in the theories of Special and General Relativity. Judging from names alone, its fate in these theories looks like a fair one. My argument will be for the conclusion that SR and GR are inconsistent with the relativity of motion in any philosophical sense.

[1] Motion is a change in the perceptions of monads (cf. chapter 1) and so a change in their attributes.

Why does the relativity of motion look like a philosopher's idea? It is not hard to see why Locke thought it was one (1924). His quite standard philosophical position is characteristically sensible, ingenious and clear. The semantic and materialist claims spring from obvious epistemological principles which Locke accepted. For him, motion and rest are just change (or lack of change) of position; a body's position is an observable relation of distance and direction between the body and certain others. Unless it relates bodies to bodies, position (and any of its derivative ideas) cannot be empirical. If not empirical, then it is not intelligible. This is an epistemological requirement for the concepts in question.

This means that, for Locke, motion is just a concept of kinematics;[1] therefore the symmetry of motion seems a necessary and obvious consequence. Here symmetry means just that any body said to be in motion relative to another may equally be regarded as at rest, the motion then being attributed to the other. It does not mean that if, say, the open sentence 'x is spinning' is satisfied by a with respect to b, then it will also be satisfied by b with respect to a. Clearly if a spins relative to b, b orbits relative to a. But in other cases, where only change of distance crops up, we get symmetry in a stronger sense. What we never get it seems is that a is in motion relative to b, but b has no possible motion relative to a: for example, b would not have a possible motion relative to a if it were discontinuous. In short, so long as motion is a simple concept of kinematics, just a matter of change in distance and direction, the relativity of motion looks a simple and obvious theorem in epistemology, backed by a theory of geometry. Relativity is a necessary truth, even an analytic one, if you persuade yourself that geometry is analytic. This is so simple and intuitive that it is difficult to see how any physical theory might upset it. But we shall see nevertheless that this intuitively appealing picture must be seriously modified in SR and GR.

A philosophical motive behind all this is ontological, pretty clearly. The Void cannot be anything. If space is an entity, on the other hand, it seems metaphysically queer in possessing none of the properties that substances are taken to possess. A great attraction in the symmetry thesis is that, if it is true, we can plainly deal with motion in a way that frees us from direct reference to spaces or space-like entities. In so far as we weaken it, we obscure the extent to which we can avoid the reference.

---

[1] Kinematics studies the geometry of motion without regard to concepts like force and mass. Dynamics includes force and mass together with the geometric ideas.

Is motion really just a concept of kinematics? It is debatable and rather complex just what is the case for classical physics.

2 Absolute motion as a kinematical idea: Newton's mechanics

For Newton, motion was a concept of kinematics: it was the concept of change of absolute place. Though Newton's best-known arguments for absolute space arise out of the part it plays in his dynamics, his fundamental idea of what absolute space really *is* is not one of dynamics. He presents a philosophical argument for there being a unique frame of reference, in the famous Scholium to the Definitions of *Principia* (Newton (1958)). It is as follows:

'As the order of the parts of time is immutable, so also is the order of the parts of space. *Suppose those parts to be moved out of their places, and they will be moved (if the expression may be allowed) out of themselves.* For times and spaces are, as it were, the places as well of themselves as of all other things. All things are placed in time as to order of succession; and in space as to order of situation. It is from their essence or nature that they are places; and that the primary places of things should be movable is absurd. These are therefore the absolute places; and translations out of those places, are the only absolute motions.'

Newton assumes that talk of places is ontically serious: there really are such things. This is not nearly so bizarre and gratuitous a bit of metaphysics as it is often made to seem. Absolute space plays an important part in Newtons mechanics, so it figures in a testable and, indeed, well confirmed theory.[1] But even outside the context of physics the idea is not beyond epistemological reach. The argument in epistemology is certainly not a proof, but it has some degree of plausibility, as I hope the reader may agree. I have no evidence that Newton considered the argument, but it is a rather obvious way to think about places, once you begin to take them seriously. A place-at-a-time is immediately accessible to observation. We can see the spatial array of objects over limited (though, maybe, large) regions just as immediately we see the objects themselves. We see objects *at* places-at-a-time. We do not discriminate the places-at-a-time by discriminating objects. For we might simply see that a certain region contains, for example, just three objects not perceptibly dissimilar from each other. I borrow the argu-

[1] See Stein (1967).

ment not from Newton, but from Goodman (Goodman (1966), pp. 195–6). Goodman's primitive notion is 'place-in-the-visual-field'. I simply replace the subjectivist strain by freezing the time to an instant. No cogency is lost by the replacement.

Now, so runs the argument, places are continuants. They are enduring entities, like sticks and stones. Something very like this idea crops up in relativity, too. Adopting a frame of reference is adopting a scheme for identifying places across time, but without taking the idea of a place and its identification with the same seriousness as Newton does. It is quite evident that being at rest in a frame of reference is the same as occupying the same place in it.

Places-at-a-time are mere temporal slices of places. The epistemological weight of the argument is that the sensory immediacy of places-at-a-time makes places ontologically serious, even though the identity of places through time is not observable. If we take places seriously, then there must be a unique, absolute answer to the question whether or not p_1 at t_1 is the same place as p_2 at t_2. For identity is not a three place relation: '$a = b$ with respect to c' is ill-formed.

This picture makes places and space accessible to observation to a modest degree and clears them of epistemological sin. I see no decisive objection to the picture, though I think I see why a relativist might prefer his own view. But Newton surely did not see this as his main argument for absolute space. Put very simply, his motive was to provide a causal explanation for motion, especially within the solar system (Newton (1958), p. 12) by a theory which would be sensitive to observation overall, rather than piece by piece. He wanted to answer the question what, if anything, was at rest in the solar system. It was surely a profound question. His answer, endorsed by centuries of later dynamicists, was that nothing was at rest, unless it were the centre of gravity of the system, which did not coincide with the centre of any celestial body. It was surely a remarkable answer.

Two ideas were crucial to the theory. One was the first law of motion:

L: Every body remains at rest or in uniform linear motion, unless acted upon by a force.

We can best read it as *an essentially geometrical statement*. Where a particle is at rest (null geometry of motion) or where its path is linear at uniform speed (simplest geometry of motion) *no causal explanation has a place*. Granted the pivotal role of the first law *L* in the machinery of Newton's mechanical explanation we cannot wonder that he should have been inclined to treat ideas of place, path and velocity with full ontological seriousness. The second crucial idea was about the notion of cause or force itself.

From an epistemologist's point of view, law L is so important because it makes the idea of force epistemically accessible. For forces (and causes generally) can now be required to meet the following conditions:

(*a*) Any force has an identifiable body (conglomeration of matter), as its source.

(*b*) The conditions under which the source body is a force centre or source are specifiable independently of any description of its effect (e.g. a glass rod is electrically charged when rubbed with a silk cloth; a body acts gravitationally under all circumstances).

(*c*) Each centre of force acts on whatever it affects in some definite, lawlike manner (e.g. by contact, inversely proportional to the square of the distance etc.).

(*d*) Causal generalisations describing forces (e.g. all like-charged bodies repel one another) are defeasible (e.g. by the clause 'unless an insulating wall is between them') but the defeating features must be causal (i.e. meet conditions (*a*)–(*c*)).

We might call this conjunction of conditions the Principle of Body Centred Forces. Plainly it is a somewhat rough characterisation of an empirical notion of force.[1] It took a genius like Newton to see that the principle and the intertial law L offered a theory of mechanics far more sensitive to observation over all, than any theory, built just on the principle of the relativity of motion, could be.

How, then, does Locke's kinematical picture of motion square with Newton's? The answer to this question is summed up by Newton as the difference between real and apparent motion. It is real motion – motion through absolute space – that Newton gives dynamical point to and in terms of which he concludes that no body in the solar system is at real rest. He recognises Locke's concept, of course, but rejects its significance.[2] Any body in the solar system may be seen as apparently at rest. But Newton argues that if we apply his mechanics to bodies in so generous a way, we cannot keep the principle of Body Centred Forces. The forces prescribed in mechanics, like inertial, centrifugal and Coriolis forces, 'come from nowhere' if we accept these bodies as at rest.

[1] But see the remarks in Feynman *et al.* (1963), Vol. i, pp. 9–4.
[2] I do not mean to suggest that Newton's ideas derive from Locke.

3 A dynamical concept of motion: classical mechanics after Newton

It is awkward, of course, that Newton constructs the idea that a body has a unique absolute real velocity, but gives this idea no further role in his explanatory scheme. This means that real velocity is not observable and has no observable consequences, as analytical empiricists tirelessly complain. But the former way of describing the awkwardness is much to be preferred, I believe. Absolute velocity explains nothing. In full, Newton offers us a concept of motion in a package with a highly successful, observationally rich and sensitive, theory. But one aspect of the concept of motion is purely idle in the theory. However, even it may seem logically essential. Acceleration is utterly crucial to mechanics; how can change of velocity be crucial unless velocity itself makes sense?

But practical difficulties in applying Newton's mechanics led workers in the field in this direction: allow yourself to choose a 'rest space' just if the first law, *L*, comes out true for the 'rest space' you pick. (Here, 'comes out true' includes its meeting the principle of Body Centred Forces.) Then you can press on using Newton's mechanics as if you had found absolute space, postponing the question whether the rest space is at real rest to another, perhaps infinitely distant, day. Of course, this means that different workers in the field may choose spaces differently. But never mind. Call this device the device of 'inertial frames' in honour of the Inertial Law, *L*, by which we choose them. Then given one such frame, there is an uncountable infinity of others: all and only those in uniform linear motion with respect to the first. Classical mechanics after Newton abandons absolute space and motion in favour of inertial frames and motion relative to them.

Using inertial frames does not mean giving up Newton's real rest in favour of his apparent rest. It is not a retreat to Locke. *It means building a dynamical condition into the very idea of rest.* Each inertial frame has to meet condition *L* in the light of the principle that forces are centred on bodies. So '*x* moves' no longer means that *x* changes position or direction relative to some body or other, but that it changes distance or direction relative to an inertial frame. The relativity of motion seemed assured directly from Locke's purely kinematic concept of motion. But the assurance is completely lost once the concept of motion turns dynamical.[1] Clearly, the symmetry thesis of relativity fails. The philo-

---

[1] Inertial frames meet the dynamical condition. Since rest, uniform motion and acceleration must all be related to them, they are all dynamical ideas. Of course, this does not mean a force is needed to act on objects at rest or moving uniformly.

sopher's picture of the relativity of motion overlooks the need to weave motion into our causal theory as the weft of the fabric of explanation.

But can a very primitive concept change just like that? If it can, just how can it? This is a knotty problem. But the questions assume that the concept of motion did change from a purely kinematical concept to a dynamical one. This is doubtful. Before Galileo, it seems clear that people generally thought (no doubt osbcurely) that space was absolute, the earth being fixed at the centre of things. If anything moved, uniformly or not, something had to cause it to move. All motion was change; all change was caused. So we might think of the Lockean concept of motion as a rather radical change in the original, primitive idea and the concept of inertial frame as a step in a conservative direction. But the ice is too thin for this kind of skating. The concept of 'concept' will not bear much weight, though I am far from thinking it entirely useless. We might say, tentatively, that to turn from Locke's picture of things to the dynamical one that 'x moves' means that 'x moves relative to an inertial frame' would not have jarred on the intuitions or the sense of fitness to the language of physicists or ordinary people. Of course it would upset empiricist philosophers who have such strong *a priori* convictions about what words can mean.

To these diffident reflections on the history of concept of motion, let me add some equally diffident remarks on what appears to be common usage in our ordinary vocabulary. Though ordinary usage permits the sentence 'x moves' to be open only at one place, it is clear that this does not point to there being a single frame of reference assumed by all standard speakers of the language. To pack the picnic lunch so that it does not move does not require us to leave the lunch behind in our brief jaunt across the terrestrial surface. But there is slim support for full relativity here. To concede that something moved carries no obligation to concede that some other thing also moved, though symmetry would appear to call for such an obligation. Where 'moves' occurs in explicitly relational contexts, it is not clearly symmetrical. It is standard to say that the car moved away from the scene of the accident but deviant, indeed bizarre, to say that the scene of the accident moved away from the car. There does seem to be a symmetry in explicitly relational sentences like 'x moved relative to y', however. I offer these few observations for what they are worth, without the suggestion that they are worth much. But ordinary usage is surely slender comfort for philosophical relativists.

The blow for Locke is not just that inertial frames are inertial, however. It is also that they are frames. Frames are not objects, but schemes or methods or procedures for identifying places across time. Despite the habits of expositors of theories (sometimes useful, sometimes merely glib) classical mechanics does not and cannot offer an assurance that any inertial frame has a 'rest' body attached to (or co-moving with) it, let alone that all frames do.

Given the law of universal gravitation, it looks rather as if classical mechanics might get closer to telling us that no body is at rest in an inertial frame. But, in fact, this is all just good or bad luck so far as the theory goes. That is, these are consequences of the laws *together with initial and boundary conditions*, not consequences of the laws proper. So in classical mechanics we lose not only the symmetry thesis of relativity. We lose the materialist thesis, too. Inertial frames are characteristically disembodied. Ontologically, they are as much or as little awful as absolute space is. Classical mechanics, in short, makes constant reference to space-like entities that endure through time.

We have a choice here, it might seem. We can have a Lockean idea of motion. We have direct epistemological access to this, but it plays no useful explanatory role. In classical mechanics, we need both the idea of forces centred in bodies and the idea of acceleration, before we can get a fully testable theory.[1] Alternatively, we can go the way of Newton making motion absolute but epistemologically remote. We get a concept of motion with a very definite explanatory role, even though uniform absolute velocity is perfectly useless. Lastly, we can relate motion to inertial frames. This does not recognise absolute velocity, but does recognise absolute changes in it. Inertial frames are epistemically reachable, if not very directly; the concept of motion loses kinematic purity but gains explanatory significance. In fact, the alternatives are not really equally open. Sermonise as one may on the virtues of directly verifiable concepts, people will just not talk in accordance with ideas of motion that take it out of the causal network. Nor are the idle features of Newtonian absolute motion endorsed by the usage of practitioners in the field. What survives is a picture of motion which offers no groundwork for a philosophical argument that motion is relative, for the relation is not symmetrical, nor is it one defined primarily to hold among material bodies.

[1] Thus I do not really see how Van Fraassen's suggestion (1970), p. 114, gives us a definite theory of mechanics.

4 Newtonian spacetime: classical mechanics as geometrical explanation

A little earlier, I made the suggestion that we might view Newton's first law *L* as mainly geometric in its sense. We hurried past this and it remained a passing thought. Any capital that can be made out of this kind of suggestion must count as real money in the bank for the purposes of this book, so we shall take a second look at it.

Suppose we regard Newton's theory not just as describing the evolution through time of mechanical systems in space, but as a theory founded in a 'space' which combines these dimensions. It will be a Newtonian 'spacetime' – a concept which we are not yet familiar with. It will be a four-dimensional differentiable manifold (the sort of thing we began with in Chapter 9) for the location of events at points – it locates point-events. A 'thing-history' is a four-dimensional material tube, or world line in spacetime made up of material point-events. Things are seen as moving apart just if their world lines incline away from one another along their temporal dimension in the direction of increasing time, and so on. The four-space has a unique partitioning into instants (classes of simultaneous events). Each instant yields by cross-section a Euclidean three-space. So spacetime is like a stack of paper sheets, each sheet being space-at-a-time. In this way spacetime differs from space which is homogeneous and isotropic like a solid block. Spacetime has a built-in breakdown into hyper-sheets, whereas a homogeneous block may be split into lower dimensional sections in a variety of ways. (See Stein (1967) and Misner *et al.* (1973), Chapter 12.)

According to our reinterpretation, both Newton and Locke agree that spacetime has this sheet-like structure. However, they disagreed as to which laws dictate how the spatial sheets are to be stacked up into a spacetime. For Locke, it is necessary only to stack them so that each particle has a continuous world line. The stacking law is merely topological. Given one sheet and a material point *p* in it, the others can be stacked in any way at all that connects *p* in all other sheets continuously with *p* in this one. For Newton, however, the stacking is more definite; in fact, it is uniquely fixed. Certain world lines must be linear and orthogonal to the space-sheets if the corresponding particles are at absolute rest. Certain other world lines must be linear and inclined at definite angles to the space-sheets, if the corresponding particles are force-free but moving at absolute uniform speed. So the sheets stack just one way.

We noted before that some of this machinery does no work. As

Huygens saw from the outset (Stein (1967), pp. 183–9) Newton's distinction between accelerated motion on the one hand and rest-and-uniform-motion on the other does not also need a distinction between rest and uniform motion. In other words, the device of relativity among inertial frames gives away no explanatory part of Newton's theory, though classical mechanics with absolute space actually does assert more than classical mechanics with limited relativity. In particular it says that absolute rest particles are possible. We are better off without this assertion. What is crucial, however, it that velocity changes should be real and hence that constant velocity (no change) should be real. But it is not necessary for this that one constant velocity be absolutely distinguished from any other. To throw out the idle assertion implies something quite definite about the stacking of space-like sheets in Newtonian spacetime.

The proper explanatory content of classical mechanics entails that the sheets be stacked so that *spacetime geodesics* are defined. Any rest-or-uniform-motion particle has a perfectly definite spacetime geometry: it is geodesical. Whereas earlier we said rest particles have null geometry and uniformly moving particles have simplest geometry, we can see that all these particles are exactly alike in a way that makes no reference to inertial frames, or coordinate systems or to anything other than the world lines themselves. Classical mechanics requires just that the spatial hyper-sheets be stacked so that the world lines of all force-free particles pierce them like straight rods or spokes. This does not define the stacking uniquely. The whole pile of sheets can be shifted by shearing motions which leave all the piercing rods straight and equidistant from one another, but alter the angles they make with the sheets they intersect. For this we must imagine the sheets not as glued to points of the rod but free to slide smoothly along them.

Now the shearing and sliding motions which the sheets are free to make correspond to the group of affine transformations of a Euclidean space. So we can see the differences between Locke, Newton and inertial-frame classical mechanics as differences in their picture of the geometry of spacetime. Locke's description of spacetime leaves everything but topology indefinite, whereas classical mechanics after Newton gives a determinate affine structure to spacetime. Geodesics are fixed in it. So the point of a spacetime perspective on classical mechanics is this: we get a very clear view of the first law, L, as a geometric statement and thus we can see the part geometric explanation, as against causal explanation, plays in classical mechanics. For the law is now obviously saying that only world lines that are not geodesical need a causal explanation for their taking the shape in spacetime that they have. The geodesical world lines are as they are because of the affine structure of spacetime. In Chapter 9 geodesics were given in spaces generally through affine structure: let a tangent (proto) vector simply 'follow its own nose' through the affinely connected manifold. This is what a particle always does unless caused to do otherwise. That explains why cause-free particles do what they do. There can be no causal explanation why force-free world lines are geodesical. That springs direct from spacetime structure – from its shape in a general (and, in this case, rather weak) sense.

By comparison, the standard frame-relative way of stating the first law *L* is complex and hard to follow. It is clear enough that the law is identifying those cases where causes are not to be looked for. It seems fairly clear that these cases are identifiable kinematically, i.e. geometrically, given the right frames. But the class of objects 'at rest *or* in uniform motion', like the class of those with 'null *or* simplest' geometry of motion is awkward and unclear. Obviously, the ideas in the disjunctive class descriptions are somehow connected. Adroit changes of frame can always transform rest into uniform motion and vice versa. But the import of the law is not so easy to penetrate as when we drop all consideration of reference frames and turn to the synoptic view in spacetime. The class of force-free world lines has a single property. Each is geodesical. Now the law stands out clearly as geometric and not causal in its explanatory import. The cause-free world lines are the simplest geometric ones. Distinctions among rest and uniformly moving particles screen our vision. When they fall away as unnecessary we see *L* for the simple truth it is.

This single advance in clarity would hardly be enough to tempt us away from classical styles of thinking in terms of things in space enduring across time. My aim in turning to spacetime here is to point out where geometric explanation began to enter physics, even if only in an imperfectly clear way, and to point out the clarity that may come by going beyond questions of motion and its relativity, to a spacetime view where the trappings of reference frames and the relativity of motion are seen as essentially beside the main theoretical point. It is tempting to go on from here and strengthen the way in which the shape of spacetime may work in explanations in classical mechanics. For it is possible to treat Newtonian gravity as a curvature in this sheet-stacked spacetime (Misner *et al.* (1973), Chapter 12). Little would be gained by probing this that we will not get later from our brief study of GR. I simply point out this possibility to the interested reader.

Let us notice, for the sake of a later contrast, that the spacetime of classical physics after Newton has only an affine and not a metric structure. We can see from the model of the stacked sheets, that any pair of points in the same sheet are a definite distance apart. We can also see that a time interval corresponds to a quite definite thickness of the stack of sheets between the times. Still, if we consider any pair of point-events, say in the history of some uniformly moving particle, we cannot specify a spacetime interval between them. In the model, a straight spoke corresponds to such a particle, but as we slide the sheets one way or another they slide along the spoke so that the length of the spoke embedded in the stack between two sheets that represent two times will vary as we shear the stack one way or another. In short the interval (length of spoke) between two point-events is not (in general) made definite by the model.

In the theory of special relativity the gain in power and scope of geometric explanation in spacetime is markedly increased. We lose the 'layer of sheets' structure of classical spacetime for something a little closer to the homogeneous solid. Let us now turn to that question.

5 *Kinematics in Special Relativity: the idea of variant properties*

Let us begin in post-Newtonian classical physics, that is mechanics without absolute space. Suppose that there is a square room at rest with respect to some inertial frame F_1. It is free from gravitational forces. Inside it a perfectly elastic projectile moves up and down bouncing off the end walls but maintaining a constant distance from the side walls, as a squash ball might almost do. In fig. 40 we imagine ourselves over-looking this room.

Now imagine a second inertial frame F_2. It contains an exactly similar room at rest in F_2 and an exactly similar ball. The F_2 room is in motion with respect to F_1 along a line parallel to the end wall c. Each ball moves at the same definite speed S *with respect to its own room*. If we describe the F_1 room and its ball with respect to F_2 we can no longer say that the ball moves to and fro along a closed path (whose end points are identical). Successive collisions of the ball with wall c occur at different places as in fig. 41. That is entailed by the room's being in linear motion with respect to F_2. The path of the ball is different relative to F_1 from its path

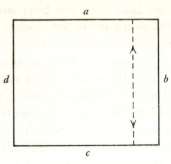

Fig. 40

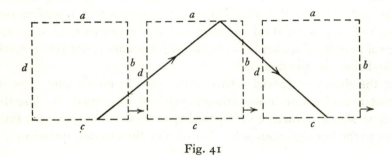

Fig. 41

relative to F_2. Let us express this in a style we have used before, in geometry. Let us say that the path varies under transformation from F_1 to F_2, or, more generally, that path is a variant under transformation. It means that we cannot ask what the path of the ball really is, absolutely. That is, a question of that absolute kind, is ill formed in this theory (though we could properly ask it in Newton's absolutist theory, even if we might not be able to answer it). The ball has a path only relatively; it is *closed* relative to F_1, *open* relative to F_2.

But it is not only with respect to these topological properties that the paths with respect to F_1 and F_2 differ. It is quite clear that the path with respect to F_2 is longer. So length of the path varies under transformation. Variation in length of path need not always coincide with transformation from one frame to another. Thus consider a frame F_3 with respect to which the F_1 room moves from right to left at the same speed as it does with respect to F_2. Its diagram of the motions of room and ball is the exact mirror image of fig. 41. Thus the length of the ball's path with respect to F_2 is the same as it is with respect to F_3 and each path is open. They differ in direction, of course. So if we define the idea of path vectorially then we can say that the path of a projectile varies with every transformation from one inertial frame to another.

The idea of quantities that are invariant under transformation from one inertial frame to another is even more important. To get some insight into this elementary notion we need to complicate our example a little and look into it more closely. Let us call on the services of the F_2 room which we mentioned (but didn't use) before. The two rooms will be in relative motion. Suppose one room moves directly over the other, so that, at one stage of the motion the rooms coincide (when viewed from above, as we suppose we are doing). At this instant let us suppose each ball collides with its lower wall c (or c' for F_2) in such a way that one collision point is directly above the other. Call these events e_1 and e_1' respectively. Then it is clear that with respect to each frame the path of its left–right moving ball is longer than the path of its 'rest' ball. The distance between walls a and c must be the same for both rooms in each system since the walls slide one above the other but the 'moving' ball's trajectory is an open path relative to each frame. Clearly, that means a longer trajectory with respect to the F that takes it as moving. Let us call the events of each ball's next (after e_1 e_1') collision with its upper wall (a, a') events e_2 and e_2', and the events of each ball's next collision with wall c events e_3 and e_3'. This is as shown in fig. 42, drawn with respect to F_1. (The dotted outlines represent room 2 at the time e_2', e_3'.) In classical mechanics it is taken that the time interval does not vary under transformation. Thus e_2 and e_2' occur simultaneously, e_3 and e_3' occur simultaneously and the e_2–e_2' pair occur midway between the occurrences of the e_1–e_1' and e_3–e_3' pairs.

Since time is to be invariant under transformation, it is clear that speed must be a variant under transformation. Trivially, velocity, as a vector idea, must vary whenever path varies: thus it could not be invariant under transformation, whatever happened to time. However, the case of speed is not trivial since speed is not a directed quantity, though velocity is a vector. But, in the case we are considering, path must vary, not just in direction, but also in length. Since time is not to vary, speed must vary in order that the longer route is traversed by the 'moving' ball in the same time. In short, it is a *direct geometrical require-ment* of the relativity of motion that path always varies under trans-formation and in such a way that length of path sometimes varies. Our example illustrates quite clearly a case in which length must vary. But it is the requirement that time does not vary which entails that speed is not invariant under transformation.

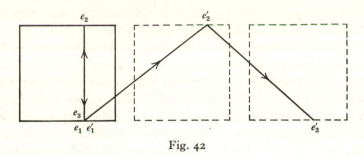

Fig. 42

This is a neat picture, easily understood and simple to generalise from. But the addition of classical electrodynamics to classical (post-Newtonian) mechanics caused some trouble. James Clerk Maxwell's theory of the electromagnetic field offers the idea of a fundamental constant speed in physics, the speed of radiation in the field. It was understood that Maxwell's equations described a medium, the ether, which carried electromagnetic energy from source charges and which reacted to any motion or vibration of a charged particle. Vibrations were carried outwards by spherical waves through the empty ether and the speed of propagation of the wave front was a fundamental constant c, the speed of light. Clearly enough, the theory also entails that the path of a light ray in a vacuum must be linear.

If we add these ideas to classical mechanics, it looks obvious that the principle of the relativity of motion will break down. Suppose we change our former example very slightly. Instead of a ball bouncing off walls of a room, think of a light pulse bouncing off mirrors. All the earlier arguments apply. The relative speed of a light ray will depend on which frame of reference we pick. Exactly one frame will both give us the value of the speed of light which Maxwell predicts and give us the relative speed of light as the same in every direction. So classical electromagnetics does more than merely pick out inertial frames from frames in general, as mechanics does. It singles out just one from all the inertial frames as the frame which tells the truth about light and electromagnetism. Furthermore, it does this kinematically. The frame is specified by a first requirement that all light paths are to be linear with respect to it. This requirement picks out, kinematically, all and only the inertial systems of mechanics. Then we add the second kinematic requirement that the radiation be propagated along all paths at the same speed. This gives us exactly one frame. Dynamical concepts are no longer needed to select the physically significant frames.

Now if this were the correct picture it should be pretty easy to find the unique frame. The earth cannot be at rest in it, except momentarily, since it moves in a curve round the sun. So it is a simple matter of performing an experiment along the lines of the example, but adapted to meet the problems of practical experimental researchers. But, as we now all know, experiments did not reveal the expected differences in relative speed of light along different paths. It was just as if the earth was always in the right frame. There seemed to be a real awkwardness at one spot in matching the new electromagnetic theory to the experimental facts.

Einstein proposed a new theory in 1905 which had both conservative and revolutionary aspects. He proposed that the principle of relativity, well entrenched in post-Newtonian classical mechanics, should be conserved. That is, that Maxwell's equations should be understood so that the laws of the field, like the laws of mechanics, would be the same when we transform from one inertial system to another. This requires that the speed of light be the same, whichever frame of reference we choose to measure it in.

Perhaps that sounds more conservative than revolutionary, but a return to our earlier example should make the impact of the proposal move obvious. We saw, in the example, that it is a very direct, simple consequence that the path of the bouncing ball must vary under a transformation. That is still true if the ball is a light pulse reflected from mirrors. It then followed directly that if the time intervals between events e_1 and e_2 were invariant, then the speed must vary relative to the inertial frame. This conditional statement is just as true as ever, of course. But now, instead of insisting that time does not vary, we insist that the speed (*when it is a light pulse*) does not vary. This means that the time interval between the two impacts (reflections) e_1 and e_2 must vary as we transform from F_1 to F_2. We need a more careful look at all this.

Take the events e_1, e_2 and e_3 and look again at fig. 42. The figure is drawn so that we are stationed in F_1. The place of e_1 and e_3 is the same – this light pulse has a closed short path relative to F_1. Suppose we have a clock at the place where e_1 and e_3 occur. Then it is a clock *at rest in F_1*. Say it shows one unit of time to have elapsed between e_1 and e_3. Obviously, the light pulse concerned in the events e'_1, e'_2, e'_3 takes a longer, open path. If the light pulse moves at the same speed over this path as its mate did over the short e_1, e_2, e_3 path, *then e'_3 occurs later than e_3 occurs.* (e_1 and e'_1 are simultaneous, since we can suppose each pulse sent out from the very same flash at the e_1 place.)

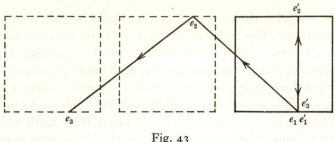

Fig. 43

But now look at fig. 43 which pictures the same events drawn so that we are stationary in F_2. Fig. 43 is the left–right reflection of fig. 42. (Again, the dotted outlines represent the different positions of the F_1 room at the time of the e_2 and e_3 events.) We get a mirror image description, too. The place of e_1' and e_3' is the same – this light pulse has a closed short path relative to F_1. Suppose we have a clock at the place where e_1' and e_3' occur. Then it is a clock at rest in F_2. It, too, will show one unit of time to have elapsed between e_1' and e_3'. (It must show one unit if the F_1 clock does. For F_1 gives a speed for light such that it moves from the top wall to the bottom along the shortest path in one unit. Since the top and bottom walls of each room slide along one another, each frame sees its shortest path as the same distance. So if each is to give the same speed to light, each clock must give the same interval for light to travel that distance.) Obviously, the light pulse concerned in the events $e_1 e_2 e_3$ takes a longer, open path. If the light pulse moves at the same speed over this path as over the short $e_1' e_2' e_3'$ path, then *e_3 occurs later than e_3' occurs.*

The immediate consequence of this is certainly dramatic: e_3' is later than e_3 relative to F_1, but e_3 is later than e_3' relative to F_2. Thus the temporal order of certain events can be completely reversed under transformation. It is important to understand that this includes the simultaneity of events. Suppose we were to consider a frame F_4 relative to which F_1 and F_2 are moving in opposite directions at the same speed. Then symmetry[1] suggests that e_2 and e_2' would be simultaneous with each other, as would e_3 and e_3'. That consequence does follow, in fact. But these relations of simultaneity for F_4 vanish under transformation to F_1 or F_2. This means that the synchrony of clocks is settled only relative to some frame or other. Clocks are synchronous just when they give the same dial reading simultaneously. If the clocks are so placed that the

[1] We are relying heavily on symmetry throughout just in adopting the relativity of inertial frames.

simultaneity of both reading, say, zero is merely relative, then they are only relatively synchronised.

The relativity of simultaneity and synchrony is not the worst fate to befall clocks in SR.[1] Moving clocks run slow. What this means is obvious enough from the argument so far. The rest clock of F_2 is in motion with respect to F_1. The F_1 clock shows 1 when e_3 occurs, but the F_2 clock does not show 1 until e_3' occurs and e_3' is later than e_3, relative to F_1. (Both clocks read 0 when e_1–e_1' occur.) So the F_2 clock runs slow relative to F_1. But, reflecting the whole argument, the F_1 clock runs slow relative to F_2. But how much a moving clock runs slow depends on the length of its associated $e_1 e_2 e_3$-type light path. That, in turn, obviously depends on how fast the clock is going relative to the inertial frame concerned. So a moving clock runs slower as a function of its speed. Results like these were revolutionary indeed.

Some loose ends need to be tidied. First, I have said nothing whatever about the role in SR of light as the fastest signal. In the context of our example, it shows itself in the assumption that the pulse of light which leaves e_1–e_1' will always catch up the F_1 or F_2 etc. clock so that there will always be some e_3–e_3' etc. event. Strictly, this is already covered by the proposal that electromagnetic laws be invariant under transformation. If some frame F_i did not get a reflection back from a light pulse, electro-magnetics could not be the same for it as for other frames.[2] Second, not every pair of events is such that their time order can be changed under transformation. If they are events in the history of the same point mass or light pulse, then no change of frame will alter their order. The same is true merely if they are so placed that they could be events in the history of one point mass or light pulse. To put it differently, let us pick any two events such that they have a purely spatial separation for some frame; that is, they occur simultaneously at different places in that frame. Then these two events will have their time relations variant under trans-formation as described just now.

Spatial intervals also vary under transformation. This is no more than a further consequence of the argument already afloat, but a new example will make it much easier to see. We need two objects, exactly similar to one another at rest in distinct inertial frames. Their being distinct simply means that they are in (uniform) motion relative to one

---

[1] Special Relativity is the physics resulting from accepting all physical laws as true relative to the special class of inertial frames.

[2] To be quite precise, I believe that SR entails that no reference frame may have a speed greater than light's with respect to any other frame. Thus SR does not make tachyons (fast particles) impossible.

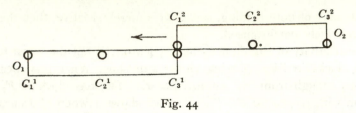

Fig. 44

another. O_1 is a rectangular object with three clocks at equal distances along an edge, so that there is a clock at each corner and one in the middle. To say O_2 is *exactly like* O_1 is to say that neither overlaps the other when they are compared for length *at relative rest*. It also means that O_1 and O_2 clocks run all at the same rate when compared *at relative rest*. The speed of O_2 with respect to F_1 (the frame identified by O_1, a rest object in it) is significantly close to the speed of light. The motion of O_2 with respect to F_1 makes it slide along O_1 so that the rows of clocks pass along one another, as fig. 44 suggests.[1]

Let us ask, first how these clocks compare with one another when they are side by side and so directly comparable. Suppose that $C_3{}^1$ and $C_1{}^2$ both read the same when adjacent, e.g. each reads o. What will the other readings be? Well, a moving clock runs slow, so we predict that $C_2{}^1$ will read a little more than $C_1{}^2$ reads at the slightly later event when $C_1{}^2$ reaches $C_2{}^1$. Let us reflect the time lag in $C_1{}^2$ by writing it in the form of an equation:
$$C_2{}^1 = C_1{}^2 + \Delta,$$
where the quantity Δ is the lag due to slowing, which would have to be added to the reading on $C_1{}^2$ to get what $C_2{}^1$ reads. Now $C_1{}^2$ moves on toward $C_1{}^1$. Naturally, it will lose a bit more time and, since the clocks are equidistant, this will be a further lag of exactly Δ. So we write a new equation to commemorate the comparison at the conjunction of $C_1{}^1$ and $C_1{}^2$
$$C_1{}^1 = C_1{}^2 + 2\Delta.$$

Now, how will the central clock on O_2, $C_2{}^2$, compare with $C_3{}^1$ when they come alongside? We can easily find this out by switching our frame of reference from F_1 to F_2 and giving the mirror image of the argument in the last paragraph. Now O_2 is at rest and O_1 passes under it (from left to right). Of course $C_1{}^2 = C_3{}^1 = o$, just as before. But now $C_3{}^1$ is a moving clock, running slow. $C_3{}^1$ will be behind $C_2{}^2$ when it reaches it. The extent to which it lags behind must be Δ, just as before, since the

[1] Fig. 44 suggests a falsehood *viz.* that O_1 and O_2 can be simply drawn as equal in length without further ado. This is the naive suggestion with which we all begin. The following argument shows what is wrong with it, but it seems best to begin by conforming to our unscrutinised assumptions.

whole situation is perfectly symmetrical. Let us write the Δ quantity always as an addition to clocks for the sake of uniform notation. This will stress the symmetry of the results, especially if we always put the Δ on the outer flank of an equation. So we write the result as below

$$\Delta + C_3{}^1 = C_2{}^2.$$

Now switch back again to F_1 and regard $C_2{}^2$ as a moving and slowing clock, which read Δ more than $C_3{}^1$ when it flashed past. By the time $C_2{}^2$ reaches $C_2{}^1$ it will have lost this little increment Δ so that $C_2{}^2$ and $C_2{}^1$ read the same as $C_2{}^2$ passes. By simple reasoning of this sort the reader can verify that we will get the results shown in the following nine-membered table.

$$
\begin{array}{lll}
C_2{}^1 = C_1{}^2 & \Delta + C_3{}^1 = C_2{}^2 & 2\Delta + C_3{}^1 = C_3{}^2 \\
C_2{}^1 = C_1{}^2 + \Delta & C_2{}^1 = C_2{}^2 & \Delta + C_2{}^1 = C_3{}^2 \\
C_1{}^1 = C_1{}^2 + 2\Delta & C_1{}^1 = C_2{}^2 + \Delta & C_1{}^1 = C_3{}^2
\end{array}
$$

Adopting F_1 as a reference frame makes it natural to read this table *down the columns* taking the $C_i{}^1$ clocks as at rest and synchronised, each $C_j{}^2$ being checked against the series of rest clocks as it goes past. The moving clocks are seen to run slow. Adopting F_2 as a reference frame makes it natural to read the table *across its rows*, taking the $C_i{}^2$ clocks at rest and synchronised, each $C_j{}^1$ being checked against a series of rest clocks. The moving clocks are seen to run slow. It is also clear that, for either frame, the moving clocks are out of synchrony with each other. It is interesting to glimpse how the symmetry of the slowing of moving clocks emerges in a new example, especially one where the very same table is used for each choice of frame.

But the main point of this example was to be a spatial one. We need to rewrite the table, supplying a bit more imaginary detail. Let us suppose that the O_1 clocks show that it takes any O_2 clock just one unit of time to pass from one O_1 clock to the next. By symmetry, in F_2, it will take any O_1 clock just one unit of time to pass from one O_2 clock to the next. So the amplified table must be the following one, as the reader should check.

$$
\begin{array}{ll}
0 = C_3{}^1; \; C_1{}^2 = 0 & 1 - \Delta = C_3{}^1; \; C_2{}^2 = 1 \\
1 = C_2{}^1; \; C_1{}^2 = 1 - \Delta & 2 - \Delta = C_2{}^1; \; C_2{}^2 = 2 - \Delta \\
2 = C_1{}^1; \; C_1{}^2 = 2 - 2\Delta & 3 - \Delta = C_1{}^1; \; C_2{}^2 = 3 - 2\Delta
\end{array}
$$

$$
\begin{array}{l}
2 - 2\Delta = C_3{}^1; \; C_3{}^2 = 2 \\
3 - 2\Delta = C_2{}^1; \; C_3{}^2 = 3 - \Delta \\
4 - 2\Delta = C_1{}^1; \; C_3{}^2 = 4 - 2\Delta
\end{array}
$$

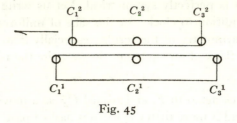

Fig. 45

Now various things about these nine results make it clear that the comparison in length between O_1 and O_2 *as they are in relative motion* cannot be the same as their comparison in length at relative rest. The reader might wish to reflect just on the import of the first column and the top row taken together. But the most graphic way of bringing the result out, perhaps, is to reflect on the central event – the conjunction of $C_2{}^1$ and $C_2{}^2$. These clocks are at the centre of each object. Take F_1 as our frame. We ask: which other clock conjunctions occur simultaneously with the $C_2{}^1$–$C_2{}^2$ conjunction? Certainly, classical ideas would lead us to expect two other simultaneous conjunctions, since when the centres of two moving bodies are adjacent, so are their ends, if the bodies are exactly similar, as we said they are. If F_1 is our frame, then we must look for other $C_i{}^1$ readings of $2 - \Delta$. But there are none. The conjunction $C_1{}^1$–$C_1{}^2$ occurs when $C_1{}^1$ reads 2, which occurs a little later than when $C_2{}^1$ reads $2 - \Delta$. So $C_1{}^2$ has not yet reached $C_1{}^1$. By contrast the $C_3{}^1$–$C_3{}^2$ conjunction occurs when $C_3{}^1$ reads $2 - 2\Delta$, which is a bit earlier than the central conjunction. So when the central clocks coincide the leading edge of O_2 is not yet at the further end of O_1, while the trailing edge of O_2 has already passed the nearer end. In a diagram, the situation of central conjunction relative to F_1 is somewhat like fig. 45.

But, once more, the same table of comparisons gives us a perfectly symmetrical reflection of the result when we choose F_2 as a frame. Now we look at the same three clock comparisons, but are guided by what $C_1{}^2$ and $C_3{}^2$ say. They are the clocks synchronised with $C_2{}^2$ relative to F_2. We find that $C_3{}^1$ does not pass $C_3{}^2$ until $C_3{}^2$ reads 2. So this conjunction occurs later than the central one, meaning that $C_3{}^1$ has yet to reach $C_3{}^2$. By contrast $C_1{}^1$ passes $C_1{}^2$ at $2 - 2\Delta$, so that $C_1{}^1$ has already passed $C_1{}^2$ when the central conjunction happens. As before, this has to mean that the moving object has been contracted (in the direction of motion). It leads to a picture like fig. 46. Again, the relativity and symmetry of the contraction emerges very clearly from different interpretations of the very same objective comparisons.

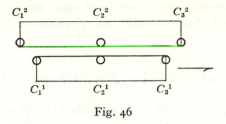

Fig. 46

Intervals of space and time vary under transformation in SR. It follows pretty obviously that this infection of relativity will spread to other physical quantities, length and time being so fundamental. Velocity is a useful example and one that will interest us later. In simple cases, where we just give the velocity of F_2 relative to F_1 and worry only about the velocity of F_1 relative to F_2, the transformation is easy and coincides with naive expectations. The velocity of F_2 with respect to F_1 is the same as that of F_1 with respect to F_2, except for reversing direction. The transformation just reverses the relevant vector sign and leaves the speed unchanged. There is a very simple symmetry. But suppose we have an object O with velocity v relative to F_2, which, in turn, has velocity u relative to F_1. Clearly the speed of O relative to F_1 is not the vector sum $u+v$. For v is a quotient of *measures in F_2* of distance and time. In F_1 these temporal measures are seen as taken by moving and therefore slow clocks. So F_1 takes the time measure to be greater than F_2 assigns it. In F_2 spatial measures are taken by rods, which are contracted (unless they are laid down in a direction lying in a plane perpendicular to the motion of F_2 relative to F_1). F_1 takes the space measure as less than F_2 assigns it. In either event, we must add u to a corrected v to get the velocity of O relative to F_1. The correction will always lessen what is added to u. The algebra of transformation for velocities is, in general, somewhat complex, therefore.

The relativity of mass and of energy will not crop up as problems for us later in the book, so I will do no more than suggest how they fit into the present still purely qualitative picture. Velocity, hence velocity changes, must always be related to some frame or other. In SR, as in classical mechanics, mass varies inversely with acceleration if the force is constant. Since acceleration varies under change of frames (addition of velocities), this strongly suggests that mass must vary too. In SR, as in classical mechanics, kinetic energy is given by the expression $\frac{1}{2}mv^2$ and momentum by the product mv, where m represents mass relative to a frame. Exploiting these facts – in the context of cases of particle collisions, for example – the relativity of mass and energy follow from what we have worked out already. (See Feynman *et al.* (1963), Vol. 1.)

6 Spacetime in Special Relativity: a geometric account of variant properties

We cannot go on without some grip on the concept of spacetime. The present picture looks very strange and unintuitive. No causal explanation is offered for why the familiar physical quantities have come unstuck and got relativised. That happened so as to maintain the speed of light (the theory of electromagnetism) as invariant under transformation. It is hard to feel that you grasp reality firmly when the matter is put like that. But the example of the nine clock comparisons already suggests something else: though space and time intervals may both vary under transformation, they do it in such a way that a very strict relationship between the two kinds of measures is kept. This, in turn, suggests that something geometrical is up. If we go straight to a naive, Euclidean spacetime picture of the example, we get very nearly what we want straight away.

Take fig. 47 as tracing the trajectories or the world lines of the six clocks we dealt with before. In any one world line, points nearer the foot of the page stand for point-events earlier than those shown by points in that line nearer the top. Since these events occur in the history of a single thing their time relations are invariant. That a line is straight means that the object portrayed is at rest in some inertial frame or other. That a pair of lines is parallel shows the corresponding objects at relative rest. That two lines are inclined to one another shows that the corresponding objects are in relative motion. The nine clock comparisons are shown by the nine intersections in the central parallelogram. In the light of what was argued before, it seems wholly reasonable to take the dotted line AB as plotting a series of events which occur simultaneously with respect to the system that takes O_1 as at rest, since AB is perpendicular to the purely time-like direction of the world lines of the O_1 clocks. So AB is a purely spatial array of simultaneous point-events with respect to F_1. Clearly enough the line XY plays a similar role for the frame F_2. We get the relativity of temporal order immediately. Events A and B are simultaneous relative to F_1, but B occurs before A relative to F_2. We get symmetrical results for events X and Y.

The figure gives us directly a change in the rate of a moving clock and a change in the length of a moving rod, though the changes are not quite the ones we are looking for. Look at the three points of incidence

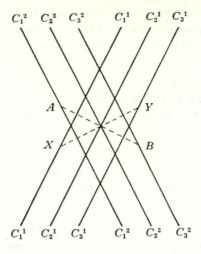

Fig. 47

along the world line C_2^1. We placed these one time unit apart, before and we might imagine them as coinciding with separate ticks of C_2^1. Suppose the same holds for points of incidence on the world line C_2^2. Clearly, C_2^2 ticks *more rapidly* than C_2^1 with respect to F_1. Consider a space-like section parallel to AB through the first (lowest) C_2^1 conjunction (tick). The first (lowest) C_2^2 conjunction (tick) lies between this section and AB, so the first C_2^2 tick occurs between the first and second C_2^1 tick. But the C_2^1 and C_2^2 ticks coincide. They are both represented by the central point of incidence. By a similar argument, we see that the third tick of C_2^2 occurs before the third tick of C_2^1 relative to F_1. The reader can easily show, now, how to read off the result that C_2^1 runs at a changed rate relative to F_2. A moving clock runs fast, therefore, according to the diagram. It is really quite obvious that the length of O_2 (the spatial interval between C_1^2 and C_3^2) is expanded in the direction of motion relative to F_2. The line segment AB is much longer than the part of it that cuts the C_1^1 and C_3^1 world lines. So moving objects expand in the direction of motion.

An increased rate for moving clocks and an expansion of moving objects are not the results we want. But the yield from even the simplest geometrical look into spacetime is strikingly high. The structure of results is exactly what we need. Is there a kind of geometry which will both yield all the structural parallels we have so far and give us a speed for light invariant under transformation and also the slowing of moving clocks and contraction of moving bodies that we do not yet have? There is such a geometry, discovered by Minkowski and incorporated into SR, as the reader is probably aware. One way to characterise Euclidean geometry is by the Pythagorean theorem: the square on the hypotenuse of any right-angled triangle equals the sum of squares on the other two sides. We can make this a general characteristic by noting that the square of the length of any interval in Euclidean space is given by the sum of the squares of the relevant coordinate differences (for Cartesian coordinates). That is we get the equation

$$\Delta s^2 = \Delta x_1{}^2 + \Delta x_2{}^2 + \ldots + \Delta x_n{}^2$$

for n-dimensional E space, where Δs is the spatial interval and the Δx_i are the coordinates differences. For Minkowski space we get the almost Euclidean expression

$$\Delta s^2 = \Delta x_1{}^2 + \Delta x_2{}^2 + \Delta_3{}^2 - \Delta t^2,$$

where Δs is the spacetime interval (and where units are chosen so as to eliminate a factor c^2 in the last term).[1]

Spacetime diagrams look like fig. 48. Take the dotted lines to represent the paths of light rays through a point event, O. T_1 and T_2 are then the world lines of two objects that meet at O. Since they are lines, each identifies a linear frame, F_1 and F_2 respectively in which the particles are at rest. T_1 and T_2, like any curves[2] that lie within the upper region between the two light lines, are time-like, so each can represent the same place at different times for its frame. The hyper surface represented by the lines L is called the light cone at O. All point-events within the upper light cone are absolutely later than O; those within the lower light cone absolutely earlier than O; those outside it – to left or right – are absolutely elsewhere, but whether earlier or later than O depends on which linear frame we relate them to. All curves like S_1 and S_2 are space-like, therefore. S_1 and S_2 represent space for their respective frames – S_1 marks out those point-events that happen at different places but at the same time, relative to F_1. S_1 and T_1 are clearly orthogonal, but S_2 and T_2 also count as orthogonal, since S_2 uniquely gives to light a constant speed in all direction, given that we chose T_2 as representing time. So rotation of coordinates in Minkowski space is more complex than in Euclidean space. Transformation from one frame to another goes according to the Lorentz transformations not the classical Euclidean or Galilean ones.

[1] For an elementary and detailed account of the parallels between Euclidean and Minkowskian geometries see Taylor & Wheeler (1963).

[2] More exactly, a curve is time-like if, at every point on it, the curve lies within the light cone at the point.

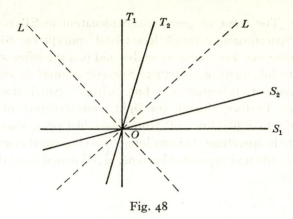

Fig. 48

The expression $\Delta s^2 = \Delta x^2 + \Delta y^2 + \Delta z^2 - \Delta t^2$ gives an invariant measure to the spacetime interval between two point-events, which does not depend on which coordinates are used. The space or the time intervals between the point-events does depend on which frame of reference we choose, as we saw before. Some aspects of this metric, obvious enough from the expression, are worth comment. If we consider light (in a vacuum) then the distance it travels in a given time is such that the sum of the squares of the spatial coordinate differences of two point-events connected (or connectible) by a light ray will always equal the square of the time interval between them.[1] So the interval Δs between any two point-events in the history of a light ray is always zero. Light paths are called null geodesics for this reason. If Δs^2 is positive, then the interval is time-like. One point-event is absolutely later than the other and the later point-event lies in the upward light cone defined at the first. If Δs^2 is negative, then the interval is space-like. One point-event is absolutely elsewhere from the other and lies outside either the upward or downward light cone defined at the other.

Of all the properties of things described in classical physics, only charge is invariant when frames are changed. These changes in properties like length and time cannot have a causal explanation, since causal relations are preserved under Lorentz transformations. The changes have a single very simple, quite intuitive, explanation geometrically once we diminish the importance of describing the world from the standpoint of reference frames and turn to a spacetime

[1] Given a choice of units which lets us drop the coefficient, c^2, of Δt^2. Clearly, this can only be so if $c = 1$. For different choices of units, the factor c^2 will still make Δs^2 equal to zero.

perspective. The scope of geometric explanation in SR is very wide therefore. Spacetime is a much less trivial concept for SR than for classical mechanics. For, now, we go beyond just the affine structure of spacetime to full metrical structure. Spacetime intervals are invariant under Lorentz transformations, but neither spatial nor temporal intervals are. Further, though the light cone structure of spacetime always marks time-like lines off from space-like ones, there is a new homogeneity in spacetime. It is no longer a well-defined stack of spatial sheets. Space and time separated become, as Minkowski said they would, mere shadows.

7 The relativity of motion in SR

What becomes of the philosophical issue of the relativity of motion in the light of all this? Strictly, what happens is that history passes the issue by. The move to spacetime and four-dimensional objects is compelling ontologically. In the classical picture, three-dimensional objects endure through time. But now these objects, which, so to speak, 'take their existence' absolutely as spatial things, have almost no determinate absolute properties as three-dimensional continuants. A thing is not a cube, for example, by itself but only for some one frame of reference. So we leave the '3 + 1 view' behind. Now it no longer matters exactly how we split spacetime into space and time. The frame of reference, as a device for *identifying places across time*, loses its point since we are concerned only with point-events and particle-trajectories. Every object has a four-velocity and a four-momentum. These are simply spacetime vectors. What worries the physicist is just whether these vectors are parallel or not, whether a particle-trajectory is linear or curved and so on. Frames of reference in spacetime have the same kind of insignificance as Cartesian coordinates in space. They are just useful but dispensable devices to help us describe spacetime and spacetime objects themselves. Questions about the relativity of motion simply lose importance, rather than gain some determinate answer.

But we can still ask these questions and the answers will be of diminished rather than of no importance. A light cone is defined at each spacetime point. This gives us regions absolutely *elsewhen* (earlier or later than the point within the downward or upward cone) and regions absolutely elsewhere (outside the cone). These facts provide a firm basis for a division of spacetime into space and time. If the spacetime velocity vectors of particles are parallel (not parallel) then this gives us back the idea of relative rest (motion). The translation rules between the two pictures are definite and cover all we want. So let us see if we can find some answers to the question of philosophical relativity.

First, the semantic thesis itself is not clearly preserved. As was just said, the four-velocity and four-momentum are spacetime vectors and do not express a relation between four-velocity and something else; nor is this point affected if we choose to describe the four-velocity in the components of a particular coordinate system or reference frame. Someone might argue that four-velocity and four-momentum are too different for this to be seen as doing serious damage to the semantic thesis. I will not debate the point, commenting only that four-velocity and four-momentum capture all the mechanical information which the classical ideas conveyed.

Next, SR gains us no new frames of reference. Classical mechanics requires us to prefer inertial frames. In SR we must prefer the class of frames which make all light paths (in a vacuum) linear and which give all light pulses a constant speed. In fact, the first condition is sufficient to identify the frames, which may be called linear frames for that reason. But though the new criterion for whether a frame is linear differs from the mechanical criterion for inertial frames, they are the very same group of frames as before. All the problems with the materialist thesis in the philosopher's idea of relative motion persist.

The symmetry thesis still has all of its earlier troubles and gains some new ones. There is an absolute motion, that of light, and this emerges as part of the kinematics of SR. Light has motion with respect to every linear frame but does not define a linear frame itself. Since the light cone plays the pivotal role, in spacetime geometry, of cleaving space-like from time-like lines, no light ray can play the role of either a time-like or a space-like line. This peculiarity of light signals in the kinematics of SR is reflected in the transformation equations among frames. Take any frame F_1, and consider another frame F_2 which moves arbitrarily close to the speed of light relative to the first. Then both spatial and temporal intervals in F_2 change under transformation to F_1 in such a way that spatial intervals approach zero and temporal ones dilate without limit as F_2 approaches the speed of light. It is a kinematical fact that space and time intervals will not go over intelligibly if we attempt to transform away the motion of light. We get nonsense in kinematics if we try. Light's motion is absolute.

Now the SR view of things makes motion look once more like a

kinematical concept, in the style of Locke. However, this offers no
return to the appealing *a priori* argument (in §1) for the philosophical
relativity of motion. Partly, this is because in losing its dynamical
element motion has become a concept of electromagnetics. But, more
importantly, it is because a significant crack has appeared in the
apparently unbreakable bridge between motion and distance-direction
change. These ideas are not so simply linked in SR as they were in
classical mechanics. There, if you knew that *a* and *b* change the distance
between them and that either is an inertial frame, then the motion
simply *was* this change of distance and direction. But things are more
complex in SR. Spatial interval, hence change of spatial interval, varies
under transformation. Spatial interval is a three-placed relation that
involves two point-events and a linear frame; so does change of spatial
interval. It differs from frame to frame in a complex way. But motion
proper is a dyadic, two-placed relation in SR between a thing and a
frame. The difference is underlined in quite different algebras for the
two concepts, as we can see in the following example.

Suppose there are three objects *x*, *y* and *z*, each co-moving with a
linear frame but none at rest with respect to any other. Suppose we ask
for the rate of distance change between *x* and *y* with respect to *z*, given
their velocities (proper motion) with respect to *z*. The answer is given
by the simple vector difference of the velocities. To take the simplest
case, if the velocities are equal in magnitude and opposite in direction,
we simply add the speed of *x* and *y* with respect to *z*. Thus, suppose the
speed of each to be $\frac{3}{4}c$, then the rate of distance change between *x* and *y*
will be $1\frac{1}{2}c$. Now ask, instead, for the velocity (proper motion) of *x*
with respect to *y*, given their velocities with respect to *z*. We need a
quite different procedure. We must use the Lorentz transformation on
x's (true) velocity with respect to *z* to calculate the rate of distance
change (apparent motion) between *x* and *z* *with respect to y*. We then add
this to *z*'s (true) velocity with respect to *y*. That is, we employ SR's
addition theorem for (true) velocities. The sum is always less than *c*. So
speed and velocity are more complex than mere distance (direction)
change.[1]

But even this limited relativity of motion in SR depends on the global
geometry of spacetime and space. SR is formulated in Minkowski
spacetime, that is, in Euclidean space. There is nothing to compel us to
make this choice of geometry either in the hard evidence for SR or in
any acceptable *a priori* principle. A choice of geometry which differs

---

[1] Compare with the discussion of velocity on p. 235.

only globally from Minkowski and Euclidean geometry produces a very strong form of absolute motion, which eclipses even Newton's ideas on the subject. Suppose we assume a spacetime of which all the spatial cross sections give us a spherical space (as in Chapter 4, §9). Then SR holds in every sufficiently small region (at all times) of this space, so that only minor modifications are needed to extend it globally and the main ideas of the theory remain intact. What will the consequences be for the relativity of motion?

We suppose that SR holds to a close approximation over any longish spacetime region, but that space has a small positive curvature, constant for each point. Spacetime has the topology of a hypercylinder, therefore. Let x and y be two objects, each co-moving with a different linear frame. Let x and y be momentarily adjacent, at which time a light pulse is omitted from x (or from each, if you prefer symmetry). This radiation will move as an expanding spherical wave front, with respect to both. (Since the speed of light is the same whatever the motion of the source, the radiations from both x and y expand together.) Eventually, the wave will reach a maximum (an 'equatorial' sphere) and begin to contract toward a point (just as a circular wave would do in the surface of a watery planet). After its collapse on the point, it would expand again in the other direction[1] (as a circular water wave would collapse on a point the polar opposite of its point of emission, and begin to move back). It will again expand to a maximum and begin to contract again *toward its point of emission*. And so on. These points are unique, of course. They are absolutely fixed across time as the point of emission and its polar point. At most one of the objects x and y can co-move with a linear system throughout and also be precisely where the light waves collapse to a point on this second occasion. So at most one will be absolutely at rest. If x is the object then its co-moving linear system is the absolute rest system, its time the absolute time, its spatial measures also absolute. But neither object need be at the emission point, obviously, in which case the absolute system has no identifying body. In fact the objects are quite superfluous to the case, and we can call just on radiation to discover absolute rest and motion for us.

But, isn't this case really at home in General rather than in Special Relativity theory? No. Though we have supposed spacetime (and space)

---

[1] If the space were elliptical instead of spherical (see Chapter 3, §10) then there would be no dual polar point. The waves would expand to a maximum, then contract toward the point of emission. This case is much harder to grasp pictorially, however, despite its simplicity in other respects.

to be non-Euclidean, I gave curvature and metric no dynamical part to play.[1] Nor has curvature been linked to the distribution of mass-energy. Nor has any attempt been made to consider frames of reference that are not linear. The characteristic features of GR were never called upon. My point was simply this: we must not suppose that SR is a step in the direction of supporting the philosopher's picture of the relativity of motion; for with no real violence to the spirit of the theory, we can embed it in a context in which it gives the clearest possible sense to a very strong form of statement that motion is absolute. The problems in GR are quite different. It is time to look into them, though we shall not probe to the bottom.

8 General Relativity and the Clock Paradox

Let us pick the relativity of motion as our clue through the labyrinth of GR. Does GR really give us both the semantic, materialist and symmetry theses and thus vindicate empiricist, positivist philosophers? One judgement, often quoted, on that question runs as follows:

> 'He [Minkowski] protested against the use of the word "relativity" to describe a theory based on an "absolute" (space-time), and, had he lived to see the general theory of relativity, I believe he would have repeated his protest in even stronger terms. However, we need not bother about the name, for the word "relativity" now means primarily Einstein's theory and only secondarily the obscure philosophy which may have suggested it originally. It is to support Minkowski's way of looking at relativity that I find myself pursuing the hard path of a missionary' (Synge (1960), p. 306).

This suggests that our clue may take us to the Minotaur rather than to safety, but at least it will raise some useful issues in a coherent way.

Perhaps the most widely, certainly the most vehemently, discussed problem in the relativity of motion is the Clock Paradox. We will not try to survey the literature on the question but point out the more striking features of GR which this intriguing problem brings before our gaze. What is the Clock Paradox?

Take a couple of twins, T_R and T_M, side by side together in a linear frame, F_R. At rest in this frame there is a line of regularly spaced marker objects, M_0, M_1, M_2, ..., M_n..., beginning where the twins are and stretching out into space as far as you like. T_M fires a rocket, which

[1] Nor do Milne & Whitrow (1949), from whom I borrow the case. See later §13.

accelerates him to a speed near light ($\sqrt{\tfrac{3}{2}}c$ is a neat example) over the space M_0–M_1. T_M 'coasts' out along the string of markers from M_1 to a point M_{n-1}–M_n. The outward leg of the journey has a symmetry for F_R: two equal bits of acceleration surround a central bit of uniform motion. Suppose T_M comes back to T_R in exactly the same way as he went out. It is a consequence of SR that a clock on T_M reads less than a clock on T_R, where 'clock' is a dummy for any temporal process at all.[1]

There is nothing inconsistent in this so far. We will be in trouble, however, if (*a*) there is some frame F_M which 'co-moves' with T_M and is acceptable as a rest frame for some theory or other, (*b*) the F_M story of the events is symmetrical with the F_R story *in the details on which the F_R calculation depends in SR*. If (*a*) and (*b*) are fulfilled, there must be a calculation relative to F_M which makes T_R read less than T_M. It can't be the case that both calculations are right.

Special relativity (SR) can be cleared of inconsistency, since it rejects the frame F_M. Given one inertial frame, the others are all (and only) those in uniform linear motion relative to the first. SR can deal with accelerating objects, of course. It would be useless dynamically if it couldn't. But it deals with them only from the standpoint of inertial frames. It would be quite obvious to T_M that his frame is not inertial because of the powerful dynamic effects of acceleration which he feels and T_R does not.

So there is a problem in GR, if anywhere. There are two elements in this theory which may suggest that it does provide a frame F_M which co-moves with twin T_M. These are the Principle of Equivalence and the Principle of General Covariance. Let us state the first of these as follows: a frame of reference which is maintained at rest in a gravitational field by a force is equivalent to a frame of reference accelerated relative to some linear frame in gravitation-free space by an equivalent force. There is a proviso: we must not try to maintain the equivalence of the frames over 'too wide' a region of spacetime. Though this is not mistaken, we will find reason later to look on it as less than the clearest statement of the Principle. Put this way, it seems to licence us to take 'accelerated' frames as at rest in gravitational fields – a significant step in the direction of relativity in the philosopher's sense. The principle of General Covariance says that all coordinate systems are equivalent for

---

[1] Notice that T_R is always in a linear frame. Without this stipulation we can make no claim as to which clock is retarded, or even that the clocks differ. If the twins were both accelerated, decelerated etc. symmetrically relative to some linear frame, the clocks would read the same.

the law-like description of physical events and processes. The laws of physics must be (and can be) formulated so as to need no reference to a special class of coordinates. This seems a yet stronger form of the relativity of motion, yet we should be cautious about identifying co-ordinate systems and frames of reference. The latter have an onto-logical significance which the former lack. Every frame is a coordinate system too, but the reverse may not be true. Let us see how we might exploit these Principles in the Clock Paradox.

Before we try out a version of the F_M story let us construct a rough catalogue of T_M's history. From the SR viewpoint of F_R we can distinguish nine stages in T_M's journey:

(1) at rest beside T_R and M_0

(2) accelerating away from T_R to M_1

(3) moving uniformly from M_1 toward M_n

(4) decelerating from M_{n-1} toward M_n

(5) at rest beside M_n

(6) accelerating away from M_n to M_{n-1}

(7) moving uniformly toward T_R

(8) decelerating across M_1–T_R

(9) at rest beside T_R and M_0

In odd numbered stages, T_M co-moves with some inertial system or other. In even numbered stages it does not. We can reasonably assume that counterparts of these nine stages will reappear in any F_M story. It will be a matter of some delicacy to work out how.

It now looks pretty obvious how to find a frame co-moving with T_M. We make up the F_M story by using the SR frames that co-move with T_M in the odd numbered stages and by using the Principle of Equivalence and the device of gravitational fields for the even numbered stages. In the even stages, T_M is poised in a gravitational field, held motionless by rocket thrust, while T_R falls freely. Since gravitational fields affect clocks, their intensity and duration may be supposed to account for the retarding of T_M's clock.

But we strike trouble with this sketch the moment we try to fill in details. Almost everything is wrong with it. Perhaps the reason why popular and semi-popular discussion of the paradox has been so persistent is that the sketch just given looks like the obvious solution, yet it breaks down so soon. The trouble is this: the intensity of the F_M 'gravitational field' and how long it lasts is a function of the rate and period of acceleration of T_M with respect to F_R. But, with respect to F_R, the rate and the period of T_R's acceleration have no appreciable effect on how slow T_M's clock will be. The slowing of T_M's clock is a function of quite different variables – the length (distance or time) of the journey and the speed at which T_M goes. T_M will accelerate and decelerate (we suppose) in just the same way whichever marker object he travels to. But his clock will be slowed more if he travels to M_{100} than if he travels only to M_{50}. So our first F_M sketch is way off the mark. It gives the clock the same retardation whatever journey is in question.

9 Time dilation: the geometry of 'slowing' clocks

A diagnosis of the trouble is that we have muddled two things. One is a classical criterion for distinguishing allowable SR frames from frames like F_M. The other thing is the set of differences between T_R and T_M that are responsible for the retarding of T_M's clock. The acceptable reference frames for SR are exactly those frames acceptable for classical mechanics. In classical mechanics we can distinguish them dynamically. They are the class of frames relative to which the Law of Inertia holds and they are called, after the law, inertial frames. In the Clock Paradox we want to say why T_M does not co-move with an SR frame and con-sequently why SR does not sustain an F_M story symmetrical with the F_R story. We invoke this dynamical criterion: F_R is inertial, F_M is not. This invocation leads us to suppose it must be the action of forces on one clock or another that accounts for the slowing. This completely mistakes the whole matter. *The phenomenon of time dilation is not a causal pheno-menon at all but a purely geometrical thing.*[1] (Even when we lay the slowing of a 'true rest' clock at the door of a real gravitational field, like the earth's, the slowing is not an outcome of forces but of the geometry of the clock's trajectory in spacetime.) That we are up against geometry, not force, is very easy to see in spacetime diagrams. The case we have been imagining does not bring in the complication of large masses and a significant curvature of spacetime. So its spacetime diagram is just fig. 49. A T_R clock measures the length (proper time) of the world line T_R and a T_M clock measures the length (proper time) of the path T_M. Obviously the paths differ in length, the Lorentzian metric of spacetime

[1] In fact, we can give a pseudo-dynamical account that sounds as if it were a causal explanation. Those who have such an account in mind will see later why I ignore it. The only point in setting it up, here, would be to dismantle it later. I regard it as teaching us nothing about the central issues. See Cochran (1960).

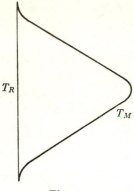

Fig. 49

making T_M shorter. Each clock measures along its path in exactly the same way as the other. In real gravitational fields we embed much the same diagram in a curved spacetime. It need not appreciably alter the picture when we do so. The real art of explaining the clock paradox is to translate the clear-cut geometrical spacetime story back into a geometrical space and time story. Whether the relativity of motion survives the translation is a delicate business which I hope to put you in a position to judge for yourselves.

Now, there is in fact no need to identify the class of frames acceptable in SR by a dynamical criterion, though this is usual. It is far more elegant and closer to the real thrust of SR to identify the frames electromagnetically as the class of frames relative to which light-trajectories are linear.[1] Now what distinguishes T_M from T_R is that at least some light paths are sometimes curved with respect to F_M, but all light paths are always linear relative to T_R.[2] Since light paths fill a pivotal place in the geometry of spacetime we can regard T_M and T_R as geometrically distinguished from one another.

This brings us to the critical point where the clock paradox bears on the philosophical thesis of the relativity of motion. For it seems to be analytically true that there is a kinematical symmetry between T_M and T_R. It seems so for Locke's[3] reasons. Firstly, T_M's account of T_R's motion must contain counterpart stages to the nine listed on p. 246. Further, it

---

[1] This has to be modified for GR, but we can leave that, for the moment.
[2] I assume the light paths are not through refracting media etc.
[3] That is, again, that motion is change of position and position must (for epistemological reasons) be defined by observable relations of distance and direction between one body and another (as group of other bodies). In short, motion is just distance (direction) change.

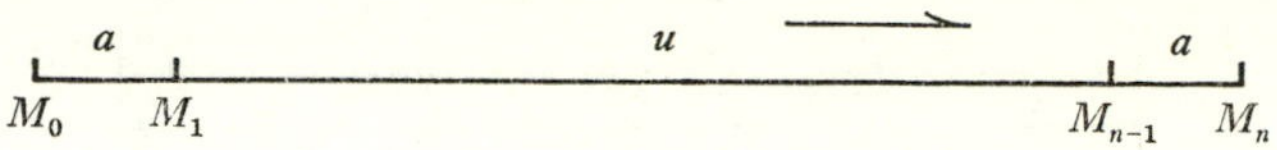

Fig. 50

seems unshakeably certain that whenever T_M is in uniform motion relative to T_R, then T_R must be in uniform motion relative to T_M; whenever T_M is accelerating relative to T_R, then T_R must be accelerating relative to T_M. This makes it impossible to see how the geometrical distinction can come to anything relevant, after all. The two accounts look as if they must be symmetrical in the geometrical variables. The whole issue looks even more paradoxical than when we began.

Here things get a bit subtle, because what seems unshakeably certain is actually senseless. The sentence 'Whenever T_M is in uniform motion relative to T_R, then T_R must be in uniform motion relative to T_M' is an ill-formed sentence of relativity theory. It is not false so much as un-grammatical. It quantifies over times without telling which frame of reference these times are times of. Or rather, it assumes that we can speak about times in an absolute way. We need to thread a verbal maze of relativity clauses to see what's wrong with this and why it matters. It will be easier followed if we first look at some spatial parallels, because we can help ourselves out of trouble with easy diagrams.

The F_R spatial picture of T_M's motion on the outward leg looks like fig. 50. That is, two intervals over which T_M accelerates surround a central interval across which he moves uniformly. But things look very different from this when we ask how T_R moves with respect to T_M. Before his rockets fire (at e_0) T_M is at rest beside T_R, in the same linear frame, F_R, in which all the marker objects are at rest. These n markers string out over some distance d, relative to this frame. But when T_M finishes firing his rockets (at e_1) he is at rest, not with respect to F_R, but with respect to a quite different linear frame. Relative to it, T_R and the string of markers are all moving uniformly (toward the left of the page, as it were) at $\sqrt{\frac{3}{2}}c$. An obvious use of the Lorentz transformation (of the sort illustrated on p. 234) shows that the row of markers, M_0–M_n, now string out over a distance of only $\frac{1}{2}d$, relative to this frame. So M_n is, at e_1, already only $\frac{1}{2}d$ from T_M with respect to his new linear frame, even though M_1 is only just passing him at this time. What must T_M conclude if he wants to combine both his linear frame before firing and his linear frame after it into one integrated non-linear frame? He must take M_n as moving (accelerating) over a $\frac{1}{2}d$ spatial interval in this first stage of the

9

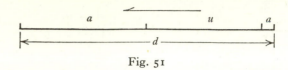

Fig. 51

journey. In short, M_n covered half the space, which has to be crossed to bring it beside T_M, by accelerated motion during the brief interval when the rockets were firing. Conversely, let us ask how far T_R is from T_M, relative to this new linear frame, just as he begins to fire his rockets for the second time (e_3). The same reasoning shows that he is now only $\frac{1}{2}d$ away beside the M_0 marker, as before. But when this second firing is finished (at e_4), T_M will again be at rest in the old linear frame F_R, together with T_R and the string of markers. So, T_R is now at a distance d from T_M. Thus T_R covered this distance $\frac{1}{2}d$ in the firing period between e_3 and e_4; that is, in the same short period begun when M_{n-1} passes him and ended when M_n comes to rest beside him. So T_M plots T_R as moving according to fig. 51 (which omits marker objects, of course). The two figures give quite asymmetrical pictures of the kinematics of the first five stages (see p. 246) of this adventure. The asymmetry is plainly geometrical.

Something like this happens with time, too. It can be illustrated in a kind of diagram, but the essential point needs to be stated verbally, as follows. A time, relative to a frame, is given by some identifying event, e.g. a green flash together with the class of events simultaneous with it *relative to the frame*. The same event can figure as identifying a time in several systems, of course, but the class of events simultaneous with it (i.e. the time) will be different for different systems. So we can't say things like 'Whenever T_M is in uniform motion relative to T_R, then T_R must be in uniform motion relative to T_M' for we are changing systems here but quantifying over times absolutely.

We need to speak more carefully, identifying particular events which each system locates in the same way, though it locates its *simultaneity class* differently. A spacetime diagram of the whole affair will focus our ideas (see fig. 52). Now, take a certain event, e.g. the hands of T_R's clock reading t. Then, relative to F_R, T_M is in uniform motion simultaneously with this event, and his clock reads t' (dotted line of simultaneity). Now, we can also say that, *relative to F_M, T_R is in uniform motion when T_M's clock reads t' (lower solid line of simultaneity). But now things get a bit tricky. If we take the time, *with respect to F_M*, at which T_R's clock reads t (upper solid line of simultaneity), that may very

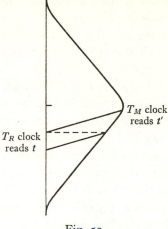

Fig. 52

well not be a time at which T'_R is in uniform motion with respect to F_M. Indeed, for the very high speed we envisaged, almost all events on T_R's clock are of this kind; that is, relative to F_R they are times at which T_M is in uniform motion, but relative to F_M they are times at which T_R is accelerating.

Now, I think this resolves the paradox. I see no way in which there is a symmetry between T_R and T_M which would suggest that we might argue that T_R's clock also runs slow. But though the paradox is behind us, the real problem about the relativity of motion has only just begun. We have lifted the lid on a result that makes it look as if we might not be able to take T_M as at rest after all.

10 *Frames of reference in flat spacetime*

What we discovered about the frame F_M is this: T_M experiences a strong gravitational force for a period of time, after which M_n is at a distance $\frac{1}{2}d$ from him, whereas it had been at distance d immediately before. This period of time is something quite independent of the distance d, so we can make it as short as we wish. Therefore, if we pick an arbitrarily distant M_n we can get it to cover an arbitrarily large distance in an arbitrarily short time relative to F_M. That is, if we take T_M as at rest and T_R as in motion then M_n (and T_R, later) must be taken to move at speeds arbitrarily far in excess of 300,000 km/sec, the usual speed assigned to light. This doesn't mean quite what it may seem to mean since, relative to F_M, the speed assigned to light in the relevant period would be far greater than 300,000 km/sec. Certainly M_n doesn't move faster than light. The problem for the idea that T_M stays in a rest frame is not that T_R outstrips light nor is it that we give the speed of light a new value. These only suggest the real stumbling block.

If we deal with matters as we have been dealing, we strike trouble with the rest points of F_M. Any rest point in a frame appears in spacetime as a *time-like line*, i.e. the world line, or trajectory of a particle which has only temporal extension and whose spatial coordinates are constant. A time-like world line always lies in the upward light cone at each of its points. Now think about the rest point of F_M at which M_n is located before the action starts. Since it is to be at rest relative to T_M we naturally want to be always time-like, yet always at distance d (relative to F_M) from T_M. The requirements are inconsistent. They do mark out a definite, continuous line in spacetime. But the trouble is that this trajectory cannot represent the same place at different times, since it is space-like in some segments. It clearly falls outside the upward light cone where it intersects it from inside and in this segment the points it connects are absolutely elsewhere. Only a consistently time-like curve can be seen as having purely temporal extension. From an absolute, that is a spacetime, viewpoint it is a creature of darkness because it intersects an upward light cone from inside, as in fig. 53. In some segments it is time-like, in other segments it is space-like. In short, we cannot take T_M as at rest *in the way we have been doing*.

The trouble lies in the attempt to project F_M out into space globally. We get plain nonsense if we try to deal with the behaviour of distant objects relative to F_M, along the rather natural lines we have envisaged. But we can provide T_M with a reference frame which will let him do physics so as to take account of what goes on nearby him at any time. There is no parallel problem with a global reference frame for T_R, however. From a spacetime point of view, we can say further that the coordinate systems defined by different 'inertial stages' of observers like T_M overlap inconsistently in spacetime regions sufficiently remote (see Misner *et al.* (1973), §6.3). But we can give T_M a purely local coordinate system defined by an object called the Fermi–Walker orthonormal tetrad (Misner (1973), §6.4–6.5). We need not concern ourselves with the details of this, which throw little light in our philosophical direction.

But we have here a clear distinction between F_M and F_R which surely compels us to choose F_R as a more correct frame of reference. Simply F_R can, while F_M cannot, project out a global space. From a philo-

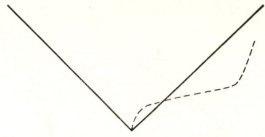

Fig. 53. The solid line shows the light cone and the dotted line
the required 'rest point' of F_M.

sophical viewpoint, what could we possibly hope to find more significant
than this as a criterion for judging that T_M cannot be taken as at rest in a
significant frame of reference? 'Only a foolish observer would try to use
his own proper reference frame far from his world line, where its grid
ceases to be orthonormal and its geodesic grid lines may even cross!'[1]

But, it is said, this does not really mean that we cannot let F_M assume
a global view of events, but only that we cannot do so in the way that
first struck us as natural. What caused the problem was our insisting
that the GR frame F_M be rigid: that is, *insisting that its rest points maintain
the same distance from one another*. The frame we need, so as to take T_M
as at rest, is one which abandons the requirement of rigidity, i.e. *that rest
points are always the same distance apart*. We need what Einstein called a
'reference mollusc' or what Møller calls a real fluid frame.[2] This is a
collection of rest points, like the points of a continuous fluid, which are
continuously but not rigidly connected with one another throughout the
fluid's history. This expanding and contracting 'fluid' can define F_M
globally. What we sacrifice is rigidity, what we gain is that every rest
point of F_M has a time-like trajectory in spacetime.

But the cost looks prodigious. It is nothing less than our deciding not
just to modify, but to jettison completely, the old Lockean idea of rest
and motion. For Locke said that a is in motion relative to b just if the
distance (or direction) between them changes. We are now on the point
of saying that distance change need not entail anything about relative
motion: that in F_M, two points p_1 and p_2 can change the distance between
them yet be at rest, in the same frame.

---

[1] Misner *et al.* (1973), p. 330. The sentence concerns the problems of an
accelerated observer in spacetime of general structure. The crossing of grid
lines requires two locations in time for the same events and is a yet more
serious absurdity than the one that arises so naturally from the Clock Paradox.

[2] Einstein (1920), Chapter 28; Møller (1952), §88, especially p. 234.

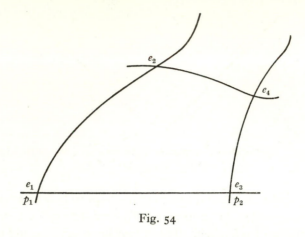

Fig. 54

This is only half the bad news. For in real fluid frames it is not merely that case that rest points p_1 and p_2 have a changing distance between them. If we take two F_M time-slices – i.e. two distinct sets of simultaneous events relative to F_M – then the amount of time between the two times is different in different places. That is, at p_1 more time elapses between e_1 and e_2 than between e_3 and e_4 at p_2 even though e_1 and e_3 are simultaneous and so are e_2 and e_4 relative to F_M. (See fig. 54.)

Here, p_1 and p_2 are more distant at time (e_1-e_2) than at time (e_3-e_4), though each is a rest point. More time elapses between e_1 and e_3 at the point p_1 than elapses between e_2 and e_4 at the point p_2, though e_1 and e_2 are simultaneous, and so are e_3 and e_4. So at different times the spatial interval between two places is different and at different places the temporal interval between two times is different.

I do not suggest that we absolutely cannot say such things. We can.[1] I argue only that, in the light of the argument so far, it gives us a powerful physical reason for rejecting the idea that T_M can be regarded as at rest and that real fluid frames are acceptable *as frames*. We can still be permissive about coordinate systems, regarded merely as labelling conventions for points in spacetime, and permissive, too, about the use of local frames of reference for accelerated observers. But the conclusion still is that GR simply does not vindicate the full relativity of motion. Even granted a relaxed view of coordinate systems and local frames, we seem to have the clearest possible reasons for preferring F_R to F_M. F_R is rigid and F_M is not. The spectacle of expanding spaces and contracting times is, so far, simply a strained artifice to pass off a coordinate system

[1] Later, we must. See §13.

as a reference frame. If we allowed ourselves free use of these artifices, there seems no reason why we should not have made free use of the artifices of centrifugal and Coriolis forces so as to accept the frames forbidden in classical mechanics. These frames were forbidden in classical physics since they did not meet the principle that every force can be identified by a body which is its source and centre (see earlier p. 218). In short, classical mechanics was not relativistic in a philo-sophical sense because only certain reference frames allowed a realistic treatment of causality. All this is still true in GR. We must again invoke inertial, centrifugal and Coriolis forces to deal with accelerating and rotating frames. These are just as fictitious as before even though they are like gravitation. Eminent writers may tell us that in GR such forces can be attributed to the gravitational action of distant matter (Møller (1952), pp. 218–20). But this is mere relativistic piety and misplaced piety at that. Distant matter was never mentioned and played no part in our account. We have discussed the whole issue in terms (virtually) of test particles with negligible mass in a spacetime of zero curvature. This is a spacetime whose average mass-energy density is zero and in which, therefore, there need be no matter at all.[1] The story is clearly a faked up business in both classical and GR cases. The classical fakery is if anything less obnoxious to intuition than the new fangled one. GR, it may seem, simply doesn't deliver the goods on behalf of philosophical relativity.

The problem of a rotating frame of reference is broadly like this, too, though it differs in detail. Suppose a disc rotates with respect to a linear frame; what happens if we try to find a frame that co-moves with the disc? Obviously, we will get the same problem as before if we look for a rigid global frame. If we took the earth as at rest, i.e. as not rotating, then related 'rest points' which are really quite close (no further away than the outer planets) would have a speed in excess of light when we refer them back to the linear frame. The idea of 'rest' clocks or rods at these 'rest points' of the rotating frame is unintelligible. This is not because we can't, even in theory, accelerate them to such speeds, but rather that they would, even so, not define real spatial or temporal intervals but only imaginary ones (in the number-theoretic senses of 'real' and 'imaginary').

[1] In fact, we are dealing here only with *curvilinear coordinate systems* in *flat spacetime*. Møller in fact never attempts description of the 'distant matter' allegedly at issue here nor is an explanation offered as to how it might produce the fields he mentioned. Since the fake gravitation field is uniform we would need an infinite mass, infinitely spread out at an infinite distance. This is cold comfort for relativistic philosophers surely.

Adolf Grünbaum mistakes this kind of objection, if I understand him. He takes it to be objectionable that, for example, the planet Neptune would need to exceed 300,000 km/sec and so 'exceed the speed of light' when we measure its motion relative to a 'stationary' earth. The objection is that this breaches a principle that SR should hold locally everywhere (Grünbaum (1964), Chapter 14, especially p. 419). But this is wrong in various ways. First, though Neptune's 'apparent velocity' exceeds 300,000 km/sec, it is not the case that the planet outstrips light (photons etc.). The 'apparent velocity' of light is also changed. All the velocities in question might be regarded as trivial or notional (Bondi (1961), p. 33). It is mere coordinate speed. Second, it is not just in rotating frames that we get 'apparent' speeds for matter in excess of 300,000 km/sec, as we saw earlier. Lastly, there is no principle which requires GR frames to preserve SR locally. Grünbaum appears to misunderstand the Principle of Equivalence. Einstein's words in 1916 (Einstein *et al.* (1923), p. 118) are these: 'For infinitely small *four-dimensional* regions the theory of relativity in the restricted sense is appropriate, *if the coordinates are suitably chosen*' [my italics]. That is clearly a quite different matter from the principle invoked by Grünbaum for which I can find no justification. The spacetime of our example is flat everywhere and plainly satisfies the Equivalence Principle. We will look more carefully into Einstein's meaning shortly.

What we can do, for the rotating coordinates, is to throw the metrical significance of our coordinate lines to the winds. Do not embark on the ultimately hopeless course[1] of taking for our coordinate time, for example, the proper time of clocks 'at rest' on the disc. Instead take the time at a point on the disc as the time shown by a circle of clocks (synchronised and at rest relative to the linear frame) revolving with respect to the disc. Møller (1952), pp. 225–6) and Grünbaum ((1964), pp. 77–80) regard this as a convenient stipulation of a time-metric for the disc and as supporting the thesis that metrics are always conventions. But it is surely clear now that we cannot do otherwise if we are to project out a global rest frame for the disc.

What does this say about GR? One might almost think that it says nothing at all, since we have not yet opened any door that leads to the real substance of GR. GR is a theory of gravitation, not of motion and rest. It says, about gravitation, that it is the curvature or non-Euclidean shape of spacetime. But we have been dealing so far with flat spacetime throughout. True, we considered non-linear coordinate systems to see

[1] Hopeless, because we cannot intelligibly project it out to Neptune.

if they were on a par with linear ones. *But this has simply nothing to do with curvature of the manifold itself.* A blackboard does not cease to be flat just because we draw some curves on it. The geometry of spacetime is still Minkowski's pseudo-Euclidean geometry, as in SR, even though our coordinate systems are not (pseudo-) Cartesian, but curvilinear.[1]

Nevertheless, the conclusions so far are important. *They show that GR fails if we try to understand it as just an attempt to make a frame of reference out of every coordinate system for spacetime.* There are admirably definite physical reasons for distinguishing and preferring linear co-ordinate systems above all others as candidates for reference frames, given a flat spacetime. If these reasons lapse when we take a wider perspective on the question, that will only serve to stress the importance of what is easy to overlook: a flat spacetime is just as surely a shaped structure and its shape just as surely important for physics, as if it were a wildly non-Euclidean manifold. But it is clear, I hope, that to under-stand GR as mainly a theory which adds new frames of reference to the spacetime of SR is to *mis*understand it. It is to cast a profoundly novel and beautiful theory in the role of a dismal failure.

11 *What GR is all about*

GR is a theory of gravitation, not of rest and motion. It says that gravitation is spacetime curvature not a force acting among bodies across space and in time. It is the *shape* of our spacetime. The bold outlines of this idea are easy to grasp. It is a geometric physical law that the trajectories of things in spacetime lie along its geodesics. More strictly, it is freely falling bodies whose world lines are geodesical. Clearly, this is an extension of Newton's first law. Bends and warps (departures from Euclidicity) in spacetime may make geodesics cross and recross; make them converge in one region though they are parallel in another; they may hold geodesics together though they are not parallel and do not converge and so on. That is, objects may sway or capture one another gravitationally in various ways.

First, look at a simple, quite unrealistic picture of a two-spacetime. Imagine a surface generally flat but with a line of hemispherical bulges in it. Each hemisphere is tangent to the next and we suppose that the line of hemispheres lies in an exclusively time-like direction. That is, if our curved surface were projected into a plane, the hemispheres would

[1] Cf. the discussion in Smart (1968), p. 238.

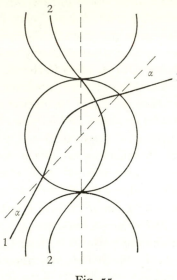

Fig. 55

project into circles and the line on which they lie would pass through the centres of all these circles.

Now, any time-like trajectory which enters a hemispherical region will leave a geodesic of the plane for a geodesic of the hemisphere in a perfectly definite way. Which great circle the trajectory lies upon will depend on the point at which the line in the plane intersects the hemisphere and on its angle with the normal to the hemisphere at that point. The great circle carries the trajectory over to the point polar opposite to the entry point and sends it out into the plane on a line which makes the same angle α as before with the normal at the exit point, and on the *same side* of it (see fig. 55, trajectory 1). Thus, in general, the trajectory leaves the region of positive curvature in a direction different from that of its point of entry. If we now consider not just one, but the line of hemispheres, it is clear that a geodesic of this variably curved surface may never escape the regions of positive curvature so as to get back into the plane. For example, if the point of exit is the point where the first hemisphere is tangent to the next, any geodesic through it will now be carried into the polar opposite point of the new hemisphere, where it is tangent to yet another succeeding hemisphere, and so on right along the line of bulges (see fig. 55, trajectory 2). Obviously, many complex variants of the simple cases of intersection with one hemisphere and complete 'capture' by the line of hemispheres is possible.

Now suppose that we are viewing this surface from outside. Between us and the surface, there is an opaque plane with a horizontal slit in it. Our perception of the variably curved surface is effectively reduced by this to a one-dimensional cross-section. Suppose that the variably curved surface is transparent, but certain 'matter' geodesics in it are solid (like wires or foil strips laminated in glass or perspex). Then suppose that the variably curved surface moves steadily up its time-like direction behind the opaque plane. Now we will see what seem to be material particles *moving about* in the linear slit which confines our vision. When a geodesic lies in the plane it will appear in the slit as a uniformly moving (or stationary) point. But when it lies on a hemisphere it will seem (in general) to accelerate. Some particles will accelerate sharply, then resume their steady course (as 1 would). Others will forever oscillate to and fro, moving rapidly in the centre of their 'orbits' and more slowly near its end points (as 2 would). Further variations are possible. Finally, let there be a geodesic, boldly marked, which lies along the line of hemispheres (that is, which passes through all their tangent points so as to coincide with the common normals at all these points as the vertical dotted line does in the figure. It will appear in the slit as a prominent rest particle which exerts a 'field of force' on surrounding particles, all nearby objects being markedly affected by it. But we know that this seeming dynamic account of our model is less profound than the purely geometric one that each particle has a world line which is always geodesical though geodesics may lie in a surface of variable curvature.

This gives a picture of the basic idea. It is a picture with some oddities, however. Our model for perception in time brings time in twice over – once as the time-like extension of the curved surface and again as the time *during which* this surface moves behind the viewing slit. This is certainly a confusion, but one that arises in a hundred other contexts in which the philosophical problem of the flow of time crops up.[1] Secondly, by making the geometry simple we made the physics hard. The 'field of force' attached to the central particle pulses. At some times it reaches out across the whole diameter of the hemisphere, but it grows and shrinks alternately, vanishing instantaneously at regular intervals. Lastly, only the central particle has an associated force field: nothing was done to provide every material trajectory with its row of bubbles. But the main point should be clear enough. It appears in *n*-dimensions that

---

[1] For an assortment of opinions on this difficult problem see the readings collected in Smart (1968) and Gale (1967).

there is a field of force with dynamical interactions pulling particles away from their simple motions. This reappears in $n+1$ dimensions as a simple geometric law for matter trajectories (each is geodesic) which holds in a geometrically complex space, the curvature of which is associated with the presence of matter. The explanatory role of dynamics (forces and causes) is taken over by geometry.

Needless to say, the Einstein theory is more complex, penetrating and ingenious than this model. But the way in which the geometry of space-time assumes the role of dynamics within space and across time, is just the same. Against this background, then, let us see if we can make the main structural features of GR stand out in strong relief.

First, there is the Principle of Equivalence. This needs to be understood as primarily a geometric principle for spacetime, rather than as an assertion about non-linear frames as rest frames. So let us look at it in this way: any very small region of spacetime can be regarded as a region of flat Minkowski spacetime. That is, very small regions of spacetime of non-vanishing curvature are equivalent to very small regions of the spacetime of SR. So that, locally, we can always find a Lorentz coordinate system. What does this mean in practical terms? For many practical purposes we regard the earth's surface as a rest frame, but we find a pervasive force field in the frame, affecting all bodies alike (so far as their trajectories are concerned) whatever their constitution or mass. The force field is gravitational. The thesis of GR is that the geodesics of this spacetime region are given by the spacetime trajectories of *freely falling* particles (and by light). Therefore trajectories of objects that maintain a fixed position relative to the earth's crust lie on *curved* (non-geodesical) paths of spacetime. Hence a force is needed to keep them on that path rather than on the geodesical one. From our worm's eye view, we ordinarily regard these forces – extended by muscles that keep us erect, by pylons that support bridges – as acting against a gravitational force. The Principle of Equivalence says that even in such regions the spacetime geometry of the freely falling geodesics is always locally Minkowskian though spacetime in the large is curved. But the Principle also figures, thereby, as a statement whose truth is a necessary condition of treating free-fall particles as geodesics. Spacetime is a patchwork of very small flat patches and its geodesics are the integrals of the Lorentz linear paths through these patches.

Of course, if the Principle of Equivalence lets us equate 'rest in a gravitational field' with curvature of a world line, it must work the trick backwards as well. So non-geodesical trajectories (accelerated particles) are just the same locally as those at rest in a field of gravity. I hope that our discussion of the Clock Paradox makes it clear why this second trick is by far the less profound and important to the structure and intention of the theory.

The Principle of General Covariance also takes on a rather different aspect when we view it from a geometric vantage point. It says that physics is really about things that are intrinsic to spacetime and independent of coordinate systems. Therefore the fundamental equations of physics must take a tensor form, since only such equations are insensitive to change of coordinates generally. This way of seeing the Principle takes it as saying that coordinate systems are *not important* at a fundamental level, rather than as saying that it is of fundamental importance that all coordinate systems should be workable reference frames. If this Principle is correct, then it means that we must deal with scalars, vectors and tensors in spacetime since, as we saw in Chapter 9, it is just these objects which are intrinsic to the manifold and through which its intrinsic geometric properties come to light. Thus in recent books on GR (Misner *et al.* (1973), Adler *et al.* (1965)) there is little or nothing about the relativity of motion. Instead, we find a great deal about the differential geometry of spacetime. We find observations about vectors and tensors in topological, affine and metric spacetime of general Riemannian structure; about geodesics, the parallel transport of vectors and the curvature of spacetime. The Lockean philosopher will find no thread of concern for the purity of analysis in terms of relations of distance and angle among material bodies to lead him through the maze. GR is simply not that kind of theory. It is precisely the *shape of spacetime*, in the sense already constructed in this book, which is the main actor in the drama of physical explanation.

12 *GR and the intrinsic geometry of spacetime*

We can build a picture of how geometry takes over dynamics by retracing the steps made in Chapter 9 to mark out how to understand a complete geometry (metric shape) as intrinsic. We began with topological structure and tangent vectors or directions at a point. We will need to begin at the same place again, because however we define spacetime curvature, we must base the idea of a geodesic and of transport of a vector and of the separation of geodesics on a primitive idea of a vector. So we need the concept of a vector in differential manifolds generally. This differs from the Euclidean concept of a vector in the way we saw (Chapter 9, §5).

What we now want is a somewhat more elaborate picture of spacetime with affine structure, roughly along the lines already sketched for the Newtonian case in §4. Every particle has, at each spacetime point along its world line, a velocity vector. Though we can express this in co-ordinates if we wish it does not describe how the particle is related to any other particle nor to any coordinate system. It simply gives the *direction in spacetime* of the particle at the point. Compare this with the view, in Chapter 9, §5 of tangent vectors as directions round a point, defined in the topology of a space. Now we go on to say that, so long as the particle is in 'free fall' its world line has the same direction from point to point, the paths being 'pointed out' by the tangent vector everywhere. That is, the world line is given by 'sliding the velocity vector along itself'. But to put this forward as a principle for all particles in free fall amounts to setting up a general affine connexion throughout spacetime. (Compare Chapter 9, §6.) It also tells us about how transport of vectors along paths quite generally changes or fails to change them.

Once we give freely falling particles and light trajectories as the geodesics of spacetime we thereby settle what is to count as the affine connexion and, therefore, the curvature (as in Chapter 9, §6). Curvature emerges in the change of direction in a spacetime vector when it is transported round a closed spacetime curve. What does this mean, however, when we are dealing with the physical behaviour of particles? In actual operations, it emerges (in principle) in experiments like this. Select a freely falling coordinate system and specify at its origin a velocity vector with purely time-like direction. It will be a particle 'at rest' in the coordinate system at that instant, therefore. Some distance away (below) (1 km, say) select another particle whose velocity vector at that spacetime point-event is also purely time-extended in this system. Let the second particle lie on a line (say the x axis of the system) which joins both particles to some centre of gravitation like the earth. The velocity vector of the second particle can be seen as a transport of the first vector to the new position. The vectors are parallel (each purely time-extended). Now consider the velocity vector of each particle one second later. That is, transport each vector along the t axis. Because the second particle is slightly nearer the centre of the earth it will no longer be at rest in the freely falling system of which the origin and t axis is given by the first particle. To compare the particle speeds again now amounts to another transport of the first vector along a space-like interval. Putting it classically, the force of gravitation would have been minutely greater on the lower particle so that they will have been accelerated to slightly different speeds (relative to the earth's crust). Putting it geometrically, though the velocity vector of the first particle (at the origin) still lies along the *t* axis, the velocity vector of the second particle *will no longer be purely time-extended* but it will have acquired a slight angle with respect to the first velocity vector (i.e. no longer be at rest in the coordinate system). This whole procedure amounts to transporting the vector at the origin in the *x* direction for 1 km then in the *t* direction for one second, plus a transport first in the *t* direction for one second then in the *x* direction 1 km. The vector is first sent round two sides of the parallelogram in spacetime (in the *xt* plane) then round the remaining two sides and compared. It is turned very, very slightly in the process, but this does measure the curvature of spacetime in this region for that coordinate plane. We could equally regard it as transporting the vector right round the parallelogram. Alternatively we could see our operation as revealing the measure of *geodesic deviation* in the region. The particles lie on geodesics that are parallel at first, but which diverge slightly a second later. The extent of this deviation of geodesics also reveals curvature. These various ways of regarding our operation add up to essentially equivalent ways of beginning to measure spacetime curvature from considerations of geodesics and an affine structure for spacetime (see Bergmann (1968), pp. 101–3; Misner *et al.* (1973), Chapter 8, Chapter 11, especially §4).

This approach, through topology and the affinity, gives us a reinterpretation of a roughly Newtonian picture of gravitation. It is more sophisticated than the picture of §4 since we have given real point to the curvature of our affine manifold. But it is still like Newtonian spacetime in that we have done nothing to rule out the shearing, sliding movements of the 'stack of papers' which was our model for affine spacetime in §4. The message of SR, however, is just that spacetime intervals (rather than spatial or temporal ones) are the invariants that give significance to SR's geometric approach to physics. To weld our theory of gravitation and our theory of SR together, we must move beyond a merely affine structure for spacetime and recognise a metric for it. Here again, the Principle of Equivalence gives us the elementary guidance. The metric of any small region of spacetime is given by the Lorentz expression for the spacetime interval in some linear coordinate system which the Principle says will always be locally available to us. The metric of spacetime in the large will be the integral over these regions. This exemplifies the procedure described in Chapter 9, §8.

The further details of the metric and its role in the dynamics of GR need not concern us. This is not a physicist's nor a mathematician's book. What I have tried to stress is that the passage of GR from topology through affine connexion to a metric is a passage through the very same stages as were argued in Chapter 9 to be intrinsic to a manifold. The same is true of spacetime. We consider velocity vectors, typically, as describing the behaviour of a particle at a given point-event in its history. But we can still speak of the whole class of spacetime vectors at a point-event whether or not they describe the trajectories of particles and whether or not there is any particle at a spacetime point. Again, a force-free particle's trajectory lies along some geodesic. But it would oversimplify things absurdly to see the particle as defining the geodesic, so that the affine structure is not intrinsic. Indeed, without the affine structure there is nothing to determine how the particle trajectory should lie. It has no antennae to tell it where other objects are, even if there were other objects and even if they acted gravitationally in a simple linear Newtonian way. It *is because spacetime has a certain shape that world lines lie as they do.* It is simply a matter of its being intelligible to speak of a curve's maintaining the same intrinsic direction from point to point.

Of course, GR does say that matter determines the shape of spacetime. But, to appeal to an earlier example (Chapter 9, §1), the fact that water has been made hot by an electric element does not mean that it is not *the water* which is hot. That the cause of curvature is extrinsic to space-time is not to say that curvature itself is extrinsic. The doctrine of GR is that matter and spacetime interact, or, better, matter and spacetime descriptions place constraints on one another. The curvature of space-time dictates how the world lines of matter must lie, but the lie of the world lines also determines how spacetime must be curved. What this is intended to express is that GR is a field theory. That is, we cannot take GR simply as a theory about the direct action of materially located matter on other distant materially located matter. The role of mass in GR is taken over by a combination of mass (or energy) and linear momentum. A tensor called the energy-stress tensor (or the matter tensor, more simply) represents the distribution and the flux of energy and linear momentum in spacetime. It is equated in Einstein's theory to a contraction (a kind of averaging out) of the curvature tensor and it is this last tensor, as we have seen, that marks out the geodesics for the freely falling particles. The geometry of spacetime plays an irreducible role.

This last point deserves less merely oracular support than I can give it here. To establish it would take us much deeper into the general physics of GR than it seems appropriate to go. I must also add that the reader would be well advised to trust himself to a guide more clearly competent than myself before he ventures into this difficult part of the forest. The case is argued in Graves (1971), Part 3 and throughout Misner *et al.* (1973). I turn to other ideas that suggest the same conclusion in the next and final section.

13 *Geometry and motions: models of GR*

Does this picture of GR require us to rethink our negative conclusions about the relativity of motion in cases like the Clock Paradox? It is not obvious how it would require this. As I already hinted before, it is hard to see that anything very central to the discussion would be changed by embedding fig. 48 in curved rather than in flat spacetime. We can still get a rigid orthonormal geodesic frame only locally. But one aspect of our criticism needs revision, though I believe this does not oblige us to retreat from ground already captured.

It is not hard to see that if we regard spacetime as being variably curved there will be examples where we cannot regard space and time themselves as rigid. This is a more serious matter than the fluidity merely of a single reference frame. The kind of thing at issue can be easily gathered from looking at the familiar variably-curved two-space of the torus in a new way. (See fig. 56.)

Regard the toral surface as a two-dimensional spacetime. It is a surface of simple variable curvature. Let *a*, *b*, *u* and *v* be segments of geodesics in the surface. Their intersections, as shown in the figure, are orthogonal so that *a* is parallel to *b* and *u* to *v*. Suppose *u* and *v* are time-like, *a* and *b* space-like. Since *a* and *b* are plainly segments of closed curves, this model represents a spacetime where space is closed and finite. The universe has a definite spatial diameter at any time. Clearly, this diameter changes. The curve on which *b* lies is greater than the one containing *a*. If we regard *b* as later than *a*, then the universe – space, that is – is expanding. There are real models of GR of a quite uncontrived kind that have this feature. For a bit more realism in our model, we must suppose a 'light cone' is defined at each point on the torus so that there is a general but definite way of distinguishing time-like from space-like curves in the manifold.

Fig. 56

Then it is obvious enough that there is no coordinate system, and so no reference frame, that enables us to escape the consequence that space itself expands and is not rigid.

Let us take a new view of the torus in which we switch space-like for time-like curves. Now we get an expanding (or contracting) time. Suppose u and v are space-like, a and b time-like segments. Then the point-events au and bu are simultaneous events (each occurring at time u). Also the point-events av and bv are simultaneous (each occurring at time v). But a clock at b measures more time (has a longer world line) between u and v than a clock at a. There is a *longer time* between two times at b than at a. Time 'runs at different rates' at different places.

Now we certainly have to reckon with such consequences in GR. Obviously, this means that we will have to look again at what we dismissed as scandalous before. We said that no global frame for twin T_M could be rigid. Both space and time might expand or contract in his frame. We refused to countenance this. But our wider view now forces us to recognise that, in some models for GR, we must countenance it. Still, much of the force of the earlier criticism can be retained, I believe. If we can select rigid frames in a given spacetime then we have the clearest and most physical of reasons for preferring them to fluid frames or reference molluscs of that universe. In some spacetimes this distinction between rigid and fluid frames will fail. But this does not deprive us of a real criterion in spacetimes where the distinction holds. We can still reject T_M's claim to be at rest when spacetime is flat.

At the end of §7 we discussed an ingenious example suggested by Milne and Whitrow. It was argued there that the spherical geometry of spacetime would allow us to distinguish absolute rest points, given the physics of SR. No final conclusion was drawn about how significant this is. Whitrow claims ((1961), pp. 221–2) that the particles must be differently related to the matter in the rest of the universe before the discrepancy in clocks attached to the particles would be consistent with

GR. Now there is reason to doubt that this is correct or even relevantly meaningful. It is certainly true that any pair of free-fall particles not at rest in the same frame will be differently related to any other matter that might be in the universe. They will differ in relative speed. But this difference is not in general what makes for differences in the rates of their clocks. The difference is simply one of their different motions with respect to each other. That is wholly a matter of the length of their world lines (which the clocks measure) and it is not entirely clear how the different orientations of these world lines with regard to those of distant matter is thought by Milne and Whitrow to be a significant feature of the physics of the situation (compare Feynman *et al.* (1963), Vol. 1, Chapter 16, §2).

More directly to the point, however, is the fact that spacetime of zero curvature can have just the features we need to construct this case in GR. If spacetime is devoid of matter then the Riemann curvature tensor vanishes and the intrinsic geometry is flat. But as we saw in Chapter 3 in the case of a cylinder, this does not mean that either space or spacetime is infinite. The topology need not be that of E_3 or E_4 any more than the cylinder needs the topology of E_2 to have an intrinsic Euclidean metric. In fact there is a valid model of GR in which space is finite, yet empty, with Euclidean metric everywhere (see Misner *et al.* (1973), §11.5). Space is closed, topologically a three-torus. The construction of the model is simple in principle. In Minkowski spacetime, pick a large cube (relative to some linear frame) say 10^{10} light years along each side. Identify opposite faces of this cube. A geodesic which meets any face simply 're-enters' from the opposite face and closes on itself.[1] We get a closed universe with finite volume, flat geometry and no matter save for our two test particles. Repeat the experiment described at the end of §7. When the particles reunite, one clock will in general be retarded with respect to the other. This gives a firm foundation for the concept of a maximal-time particle (a minimally retarded clock). That will be a particle at absolute rest. We need not, indeed cannot, take this to be a question of relationships it bears to other matter. Absolute rest is here fixed by the shape of spacetime. It is purely and simply a matter of geometry.

Still, it might be thought, GR is idling here since the real machinery of the theory relates matter to spacetime geometry. Let us therefore look

---

[1] This amounts to bending the cube without stretching it. This is a possible posture for the cube embedded in a space of sufficiently numerous dimensions so it is a possible intrinsic shape for the cube.

at a sketch of a quite realistic GR model of the actual universe (Misner *et al.* (1973), Chapter 27). Astronomy suggests that, in the very large, the distribution of mass energy is everywhere homogeneous and iso-tropic. It is a standard boundary condition on solutions in Einstein's field equation that space be finite (Misner *et al.* (1973), p. 704). Under these conditions the universe expands in such a way that the galaxies all 'move away from one another', any galaxy seeming to be the centre from which the others move, the 'speed of recession' being greater for the more distant galaxies. Astronomical observations confirm that the galactic motions appear consistent with this. This situation is possible *only if space is itself expanding*. No motion of the galaxies in a static finite space could result in each being a centre from which the others recede. We must look at the situation in the light of a familiar model of pieces of confetti stuck to points on the surface of a balloon (Misner *et al.* (1973), p. 719). If the balloon is inflated, the distance between the pieces of confetti changes in the way required, but the pieces remain stuck to the same point all the time.

In short, we have here a collapse of purely relationist pictures of motion and a decisive role for the shape of space and of spacetime in our explanation of what is going on. Our criterion for a particle's being at absolute rest in a homogeneous, isotropic and expanding universe can be exactly the same as before. Light pulses emitted from it eventually collapse on it unless blocked. It is also a maximal-time particle. If space is expanding and the universe remains homogeneous and isotropic, every particle which is a 'centre of expansion' will qualify as absolutely at rest. *Yet the distance between all these rest particles is constantly changing.* It was just the equating of motion with change of distance or direction that made Locke's version of the relativity of motion look so inescapable. But that equation now seems utterly subverted. What subverts it is the global character of spacetime – its shape, quite simply. The concepts of GR are absolutist. It vindicates the explanatory power of geometry at the expense of the older empiricist ideal of the relativity of motion.

Let us look briefly back over our whole journey. I argued that relationist accounts give no plausible intuitive picture of what it means to say that space exists. Space emerged as a particular. To understand how it is a particular we need to grasp space as shaped. I introduced this idea by getting it to do some work in a context as familiar to us as our own two hands. The shape of space explains enantiomorphism. Then we dug deeper by contrasting spaces of different shapes and by seeing shape as intrinsic to space in a general perspective of geometries. From the abstract we turned, in Chapter 4, to concrete and visualisable instances of different spaces. Here we stumbled into the snares of conventionalism. Some account was given of what methods a conventionalist must use and defend. We looked long and critically at these methods in Chapter 6 and turned aside from them. I argued in Chapters 7 and 8 that conventionalist attacks upon topological and metrical structures would not succeed even if we were to grant the propriety of the methodology which directs these attacks. I went on to explain how we could understand topology, affine and metrical structure as each intrinsic to a space. This explanation took the rejection of conventionalism as already granted, of course. Lastly, we turned to the physical world to discover an important and growing part for geometrical explanation to play there. The upshot is that geometrical explanation has a new stature in GR as an intellectual tool that we must reckon with. We cannot reject this new paradigm of explanation as a metaphysician's toy. It has done too much and is too well-entrenched for that. A proper attitude to the idea of the shape of a space is, not to evade it, but to understand it.

Bibliography

Abbott, Edwin. *Flatland: A Romance of Many Dimensions*. Blackwell, 1932.

Adler, Claire Fisher. *Modern Geometry: An Integrated First Course*. McGraw-Hill, 1958.

Adler, R., Bazin, M. and Schieffer, M. *Introduction to General Theory of Relativity*. McGraw-Hill, 1965.

Aleksandrov, A. D., Kolmogorov, A. N. and Laurent'ev, M. A. (eds.). *Mathematics: Its Content, Methods and Meaning*, Vol. 1. Trans. S. H. Gould and T. Bartha. M.I.T. Press, 1964.

Alexander, H. G. (ed.). *The Leibniz–Clarke Correspondence*. Manchester University Press, 1956.

Angell, R. B. 'The Geometry of Visibles'. *Noûs* (1974), 87–117.

Ayer, A. J. *Language, Truth and Logic*. Gollancz, 1936.

  Problem of Knowledge. Macmillan, 1956.

Bennett, Jonathan. *Kant's Analytic*. C.U.P. 1967.

  'The Difference Between Right and Left'. *American Philosophical Quarterly* 7, 3 (July 1970), 175–91.

Bergmann, Peter G. *The Riddle of Gravitation*. Scribner, 1968.

Bondi, H. *Cosmology*. C.U.P. 1961.

Bonola, Roberto. *Non-Euclidean Geometry*. Dover, 1955.

Brown, Ronald. *Elements of Topology*. McGraw-Hill, 1968.

Campbell, N.R. *Elements of Physics*. C.U.P. 1920.

Carnap, Rudolf. *Der Raum*. Reuther und Reichard, 1922.

  The Logical Structure of the World. University of California, 1967.

Chiu, H. Y. and Hoffman, W. F. *Gravitation and Relativity*. Benjamin, 1964.

Clifford, W. K. *The Common Sense of the Exact Sciences*. Dover, 1955.

Cochran, W. 'The Clock Paradox', in A. de Beer (ed.), *Vistas in Astronomy*, III. Pergamon Press, 1960.

Courant, Richard. 'Mathematics in the Modern World'. *Scientific American* (September 1964). Reprinted in Morris Kline (ed.), *Mathematics in the Modern World: Readings from Scientific American*. W. H. Freeman, 1968, pp. 19–27.

Courant, Richard and Robbins, Herbert. *What is Mathematics?* O.U.P. 1941.

Cresswell, M. J. 'The World is Everything that is the Case'. *Australasian Journal of Philosophy* 50 (1972).

d'Abro, A. *The Evolution of Scientific Thought from Newton to Einstein*. Dover, 1950.

Daniels, Norman. 'Thomas Reid's Discovery of a Non-Euclidean Geometry'. *Philosophy of Science* **39** (1972), 219–34.

Descartes, René. *Philosophical Writings*. Trans. and ed. G. E. M. Anscombe and P. T. Geach. Nelson, 1954.

Devitt, Michael. 'Singular Terms'. *The Journal of Philosophy* 71 (1974), 183–205.

Earman, John. 'Kant, Incongruous Counterparts and the Nature of Space and Space-Time'. *Ratio* 13, 1 (June 1971), 1–18.

'Are Spatial and Temporal Congruence Conventional?' (forthcoming).

Eddington, Sir Arthur. *Space, Time and Gravitation: an Outline of the General Theory of Relativity*. Harper Torchbooks, 1959.

Einstein, Albert. *Relativity: the Special and General Theories: a Popular Exposition*. Methuen, 1920.

Einstein, Albert, Lorentz, H. A., Minkowski, H. and Weyl, H. *The Principle of Relativity: a Collection of Original Memoirs on the Special and General Theory of Relativity*. Dover, 1923.

Euclid. *Elements*, Books 1–6, 11 and 12. Ed. I. Todhunter. Everyman, 1961.

Feynman, Richard P., Leighton, Robert B. and Sands, Matthew (eds.). *The Feynman Lectures on Physics*. 3 vols. Addison-Wesley, 1963.

Fine, A. 'Reflections on a Relational Theory of Space'. *Synthese* 22 (1971).

Frege, Gottlob. *Foundations of Arithmetic*. Trans. J. L. Austin. Blackwell, 1950.

Gale, R. M. (ed.). *The Philosophy of Time*. Anchor, 1967.

Galileo. *Dialogues Concerning Two New Sciences*. Macmillan, 1933.

Gamow, George. *Mr Tompkins in Wonderland*. C.U.P. 1939.

Gardner, Martin. *The Ambidextrous Universe*. Pelican, 1964.

Glymour, Clark. 'Topology, Cosmology and Convention'. *Synthese* 24 (July–August 1972), 195–218.

Goodman, Nelson. *The Structure of Appearance*. 2nd ed. Bobbs-Merrill, 1966.

Graves, J. C. *The Conceptual Foundations of Contemporary Relativity Theory*. M.I.T. 1971.

Grünbaum, Adolf. *Philosophical Problems of Space and Time*. Routledge and Kegan Paul, 1964.

'The Denial of Absolute Space and the Hypothesis of a Universal Nocturnal Expansion: a Rejoinder to George Schlesinger'. *Australasian Journal of Philosophy* 45, 1 (May 1967).

Geometry and Chronometry in Philosophical Perspective. University of Minnesota Press, 1968.

'Space, Time and Falsifiability'. *Philosophy of Science* **37**, 4 (December 1970), 469–588.

Helmholtz, H. von. 'On the Origin and Significance of Geometrical Axioms.' In James R. Newman (ed.), *The World of Mathematics*. Allen and Unwin, 1960, pp. 647–68.

Hilbert, D. and Cohn-Vossen, S. *Geometry and the Imagination*. Chelsea, 1952.

Hopkins, James. 'Visual Geometry'. *Philosophical Review* 82 (January 1973).

Hospers, John. *An Introduction to Philosophical Analysis*. 2nd ed. Routledge and Kegan Paul, 1970.

Johnson, W. E. *Logic*. Dover, 1964.

Kant, Immanuel, *Prolegomena to Any Future Metaphysics*. Manchester University Press, 1953.

The Critique of Pure Reason. Trans. Norman Kemp Smith. Macmillan, 1961.

Kant: Selected Pre-Critical Writings. Trans. G. B. Kerferd and D. E. Walford. Manchester University Press, 1968.

Klein, Felix. *Elementary Mathematics from an Advanced Standpoint*, Vol. 2, *Geometry*. Dover, 1939.

Leibniz, G. W. *The Monadology and Other Philosophical Writings*. Trans.
  R. Latta. Clarendon Press, 1898.
Lieber, Lillian. *The Einstein Theory of Relativity*. Holt, Reinhart and Winston,
  1936.
Linsky, L. (ed.). *Reference and Modality*. O.U.P. 1971.
Locke, John. *An Essay Concerning Human Understanding*. Ed. A. S. Pringle-
  Patterson. Book 2, Chapters 13 and 14. Clarendon Press, 1924.
Milne, E. A. and Whitrow, J. G. 'On the So-called Clock Paradox of Special
  Relativity'. *Philosophical Magazine* 40 (1949), 1244–49.
Misner, Charles W., Thorne, Kip and Wheeler, J. M. *Gravitation*. Freeman,
  1973.
Møller, C. *Theory of Relativity*. O.U.P. 1952.
Nerlich, Graham. 'A Scrutiny of Reference'. *Canadian Journal of Philosophy* 1
  (1972).
  'Pragmatically Necessary Statements.' *Noûs* 7, 3 (September 1973), 247–68.
Newton, Isaac. *Principia Mathematica*. Ed. Cajori. University of California
  Press, 1958.
Overseth, Oliver E. 'Experiments in Time Reversal.' *Scientific American*
  (October 1969), 88–101.
Patterson, E. M. *Topology*. Oliver and Boyd; Interscience, 1956.
Piaget, Jean and Inhelder, B. *The Child's Conception of Space*. Trans. F. J.
  Langdon and J. L. Lunzer. Routledge and Kegan Paul, 1956.
Poincaré, H. 'La Mesure du Temps'. *Revue de Métaphysique et de Morale* 6
  (1898), 1–13.
  'Des Fondements de la Géométrie, à propos d'un Livre de M. Russell.'
  Revue de Métaphysique et de Morale 7 (1899), 251–79.
  'Sur les Principes de la Géométrie, Réponse à M. Russell.' *Revue de
  Métaphysique et de Morale* 8 (1900), 73–86.
  Dernières Pensées. Flammarion, 1913.
  The Foundations of Science. Science Press, 1946.
Popper, Sir Karl. *The Logic of Scientific Discovery*. Hutchinson, 1959.
  Conjectures and Refutations. Routledge and Kegan Paul 1963.
Quine, W. V. *From a Logical Point of View*. Harvard University Press, 1953.
  Word and Object. M.I.T. Press, 1960.
  The Ways of Paradox. Random House, 1961.
  Set Theory and Its Logic. Harvard Belknap Press, 1963.
  Ontological Relativity and Other Essays. Columbia University Press, 1969.
Reichenbach, Hans. *The Philosophy of Space and Time*. Dover, 1958.
Reid, Thomas *An Inquiry into the Human Mind and the Principles of Common
  Sense*. In W. Hamilton (ed.), *The Works of Thomas Reid*. McLaughlin,
  Stewart and Co., 1847.
Remnant, Peter. 'Incongruous Counterparts and Absolute Space.' *Mind* 62,
  287 (July 1963), 393–9.
Riemann, B. 'On the Hypotheses which Lie at the Foundations of Geometry.'
  In David E. Smith (ed.), *A Sourcebook of Mathematics*, Vol. 2. McGraw,
  1929, pp. 411–25.
Rock, Irving, *The Nature of Perceptual Adaptation*. Basic Books, 1966.

Rogers, E. M. *Physics for the Enquiring Mind.* O.U.P. 1960.

Russell, B. 'Sur les Axiomes de la Géométrie'. *Revue de Métaphysique et de Morale* 7 (1899), 684–707.

Schilpp, P. A. (ed.). *Albert Einstein: Philosopher-Scientist.* Library of Living Philosophers. Open Court, 1949.

The Philosophy of Rudolf Carnap. Open Court, 1963.

Schlesinger, G. 'What Does the Denial of Absolute Space Mean?' *Australasian Journal of Philosophy* 45, 1 (May 1967).

Schrödinger, E. *Space–time Structures.* C.U.P. 1963.

Segall, Marshall H., Campbell, Donald T. and Henshaw, Melville, J. 'Cultural Differences in the Perception of Geometric Illusions.' *Science* **139** (1963), 769–71.

Sexl, Roman. 'Universal Conventionalism and Space-Time.' *General Relativity and Gravitation* **1** (1970), 159–80.

Sklar, L. 'Incongruous Counterparts, Intrinsic Features and the Substantiviality of Space.' *The Journal of Philosophy* 71 (1974).

Space, Time and Space-Time. University of California Press, 1975.

Smart, J. J. C. 'Measurement.' *Australasian Journal of Philosophy* 37 (1959), 1–13.

Between Science and Philosophy. Random House, 1968.

Somerville, D. M. Y. *The Elements of Non-Euclidean Geometry.* Dover, 1958.

Stein, H. 'Newtonian Space-Time.' *Texas Quarterly* **10** (1967), 174–200.

Strawson, P. F. *Individuals.* Methuen, 1959.

The Bounds of Sense. Methuen, 1966.

Synge, J. L. *Relativity, the General Theory.* North-Holland, 1960.

Swinburne, Richard. *Space and Time.* Macmillan, 1968.

Review of Adolf Grünbaum: *Geometry and Chronometry in Philosophical Perspective. British Journal for the Philosophy of Science* 21 (1970), 308–11.

Tarski, Alfred. *Logic, Semantics, Metamathematics: Papers from 1923 to 1938.* Trans. J. H. Woodger. Clarendon Press, 1956.

Taylor, Edwin F. and Wheeler, John A. *Spacetime Physics.* W. H. Freeman, 1963.

Van Fraassen, Bas C. 'On Massey's Explication of Grünbaum's Conception of Metric.' *Philosophy of Science* **36** (1969), 346–53.

Introduction to the Philosophy of Time and Space. Random House, 1970.

Von Senden, M. *Space and Sight.* Methuen, 1960.

Weyl, Hermann. *Space–Time–Matter.* Trans. H. L. Brose. Dover, 1922.

Whitrow, J. G. *The Natural Philosophy of Time.* Harper and Row, 1961.

Index